T5-DHG-032

IT'S A HARD LIFE
IT'S A *good* LIFE

IT'S A HARD LIFE, IT'S A GOOD LIFE
COPYRIGHT © 2015 BY META WATTS

ALL RIGHTS RESERVED. NO PART OF THIS BOOK MAY BE USED OR REPRODUCED IN ANY FORM, ELECTRONIC OR MECHANICAL, INCLUDING PHOTOCOPYING, RECORDING, OR SCANNING INTO ANY INFORMATION STORAGE AND RETRIEVAL SYSTEM, WITHOUT WRITTEN PERMISSION FROM THE AUTHOR EXCEPT IN THE CASE OF BRIEF QUOTATION EMBODIED IN CRITICAL ARTICLES AND REVIEWS.

COVER PHOTO: GRAULICH'S DUTCH BARN
PHOTOGRAPHY BY BARBARA NARK
COVER AND BOOK DESIGN BY JESSIKA HAZELTON

PRINTED IN THE UNITED STATES OF AMERICA

THE TROY BOOK MAKERS · TROY, NEW YORK · THETROYBOOKMAKERS.COM

TO ORDER ADDITIONAL COPIES OF THIS TITLE,
CONTACT YOUR FAVORITE LOCAL BOOKSTORE
OR VISIT WWW.TBMBOOKS.COM

ISBN: 978-1-61468-305-6

IT'S A HARD LIFE
IT'S A *good* LIFE

STORIES OF TWELVE
SCHOHARIE COUNTY DAIRY FARMERS
TOLD IN THEIR OWN WORDS

WRITTEN BY **META WATTS**
PHOTOS BY **BARBARA NARK**

ACKNOWLEDGEMENTS

This book would not have been written if I did not have lots of help. <u>Lots</u> of it.

Thank you first to my husband, George, for his constant love, support, patience, understanding, and wisdom, and for always, always, being there for me.

Thank you to the dedication and skill of my photographer and friend Barbara Nark, who can see light and space and expression in ways I never could. Her photographs add so much to the stories. All photos are Barbara's unless otherwise noted.

Thank you to my dear friends Paul Lamar and Elena Lobatto, who read the stories and gave me invaluable feedback. To Paul, for his unwavering encouragement, helping to keep the stories authentic, and for tirelessly correcting my terrible punctuation and misuse of apostrophes. To Elena, for her constant enthusiasm, and for providing an invaluable sense of continuity and how a story should flow.

Thank you to my daughter, Heidi Andrade, who was there with technical advice on so many levels, from computers and writing to syntax and content.

Thank you to my granddaughter, Adrienne Watts, who helped with my book proposal, and gave me such great advice about marketing and PR.

Thank you to Kim Rose, for patiently - so *very* patiently - teaching me how to use my new computer in order to complete this undertaking.

Thank you to Todd Fiske, formerly of the Northeast Center for Agricultural & Occupational Health, for his vast knowledge of marketing venues and never losing his enthusiasm for the project even after moving to Tennessee.

Thank you to Debbie Tilison, owner of Race Printing, of Cobleskill, for providing the bookmarks for each copy of the book.

Thank you to Jon Lobatto for his generous financial contribution to help with the publishing costs.

Thank you to Roberta and Jim Brooks, owners of Cat Nap Books in Cobleskill, for their kindness and much good advice.

Thank you to my children, Heidi, Michael, and David, for always believing in me and letting me know I could do this.

Oh, and thank you Jessika, from Troy Book Makers (not the betting kind) without whose knowledge, skill, kindness, and amazing patience I could never have finished this project.

I also thank my parents, Peter and Engel Petersen, who raised me on a farm and taught me the work ethic involved, and to love and respect the land, as well as the people who work it.

And last, but definitely not least, thank you to all those dairy farmers who gave generously of their time amidst their impossible schedules, and for opening their hearts and sharing their memories. This book is dedicated to them.

AUTHOR'S NOTE

Dairy farms are disappearing at an alarming rate. It makes my heart hurt. So I decided to collect and write down some farmers' stories for posterity.

These are the farmers' own stories, and as they unfold I hope the reader can feel the families' frustrations and joys, and learn what drives these grass-roots people to continue through hardships of all kinds. Perhaps obtaining a deeper understanding of a dairy farmer's life will bring renewed respect and honor to those who provide us with our daily milk, cheese, yogurt, butter, and ice cream. Really, who would want to live in a world without ice cream?

Farmers love the land. They know how hard the work is but choose it anyway. They accept that the work needs to be done seven days a week, three hundred and sixty five days a year, and they choose to do it anyway. "It's a Hard Life, It's a Good Life," is a statement of faith from more than one farmer I've interviewed.

The willingness of these people to share their stories tells me that it is vital to record and preserve a way of life that is not only rapidly disappearing, but is often misunderstood by the non-farming public. I have tried to use direct quotations whenever possible. Nevertheless, in the course of many interviews some memories were triggered by other memories and the conversations tended to wander. Despite having to rearrange the order of some conversations and combine others, I have stayed true to what the farmers said.

I conducted my first interview in 2008; now it is 2015. That is a long time for a book to unfold, but life sometimes got in the way of my plans. Though some of the farmers' circumstances have changed during that time, I have chosen to let the stories stand - with updates where provided - so the essence of each farmer's life would remain clear.

TABLE OF CONTENTS

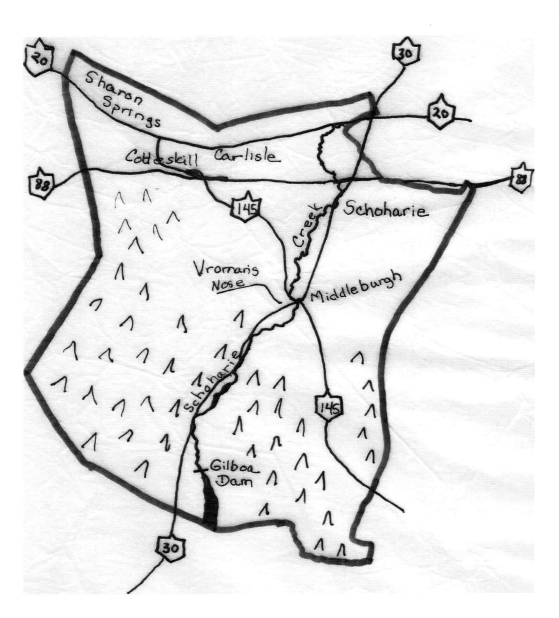

SCHOHARIE COUNTY, N.Y.

(NOT TO SCALE)

INTRODUCTION

Schoharie County is located in central New York State, south of the Mohawk River and west of the mighty Hudson. It is on the very western edge of what is considered the Capital District of New York State, Albany being about forty miles to the east. This beautiful fertile area, known as The Breadbasket of the Revolutionary War, is filled with a variety of natural wonders. The southern part of the county is in the Catskill Mountains, the northern part more rolling hills, valleys, and wide, flat fields. There are underground caves to explore, steep escarpments to climb, and the northern flowing Schoharie Creek winding through the fruitful fields. The county's most prominent feature is called Vroman's Nose, which rises high above the valley flats just west of the village of Middleburgh.

Despite its proximity to Albany, Schoharie County is largely rural; its primary industry is agriculture. According to the census of 2010, the population was 32,749, making it the fifth least populated county in New York. That same census states that the median income for a family was $43,118.

These are the stories of some of those people who have chosen to make their living in agriculture and the reasons why they choose to continue, even though that choice means they will struggle financially throughout their lives. As you read these stories you will discover that these amazing, resilient people know there are other, more important things in life to value.

EDDIE, ED, TERRY, AND GEORGE

TAFFY

PHOTOS PROVIDED BY THE BRADT FAMILY

ED AND TERRY BRADT

Cobleskill

INTERVIEWED ON MARCH 20, 2009,
ALONG WITH SON, EDDIE, AND PARTNER GEORGE MADL.

Ed and Terry Bradt share a lot in their lives - community, family life, their love for each other and for dairy farming, *and* their birthday. Though three years apart, they were both born in Cobleskill Hospital on December 13th, and were even attended to by the same nurse.

Both Ed and Terry come from a long line of farming families in the county. George Madl, Terry's nephew who also grew up on a farm, has joined the Bradts as a partner. Ed and Terry's son, Eddie, farms with them while attending SUNY Cobleskill. In addition to milking the family herd every day, Eddie also milks at the campus dairy barn.

"You've got to love farming to do it that much," he says.

Terry's grandfather and parents, as well as George's parents, worked the Madl farm in Cobleskill. Ed's parents had the old Kniskern Farm in Carlisle. Since their marriage in 1987, Ed and Terry have always worked on farms except during a short time when Ed worked in the local lace factory.

"That didn't last long, because I didn't like being inside all the time. That's not for me," he says.

They are renting the place of their dreams, a farm most local people still call The Nellie Gordon Farm, on State Route 145 north of the village of Cobleskill. When they were engaged they would drive past here and fantasize about it someday being their own. Ed reminisces:

"When I was a kid the Nellie Gordon Farm was state of the art. When we visited here it was like stepping into the future of farming. They had the first milking parlor[1] in the area."

---

1 However, he still prefers the tie-stall method of milking. "That way if a cow is off her feed you can tell right away – in the parlor, it can take two or three days to notice, and by then you've lost a lot of milk."

The Nellie Gordon Farm was started sometime in the 1800s and in its prime was a showcase place.

"Nellie was known for her hollyhocks and rhubarb. They filled the whole front yard," Terry says. "The original icehouse is still in good shape - Nellie had her tax office in it and now I have my ceramics shop there. The hops barn across the street is also in good shape - there aren't too many of those left."

The Bradts' enjoy living and working on the farm because, in Ed's words, "I like being my own boss. If I want to take a nap, I take a nap - nobody can tell me I can't. When I want to go fishing, I can. Of course, I don't get to go fishing much 'cause I don't have the time, but it isn't because somebody else is telling me I can't."

Terry loves the farm life for her family because they are always together. Her goal is to have a nice, neat, good-looking farm and she works hard and successfully at this goal. She loves to have both family and friends around.

"It's a great hearty atmosphere for kids," she says.

She likes that there are always things for a kid to do on a farm, and that they learn to help their neighbors and family.

"Kids learn responsibility and respect. That's the best thing I can say about the farm. We've raised really good kids - I've raised more than just Eddie, here. You don't meet many farm kids that aren't good kids. We like that we don't have to wait for the weekend to have fun. As long as the chores are done and there's no trouble with equipment or the cows, we can go fishing, snowmobiling, hunting, whatever. We can have a family fun day every day. When we're getting ready for the Sunshine Fair[2] and are scrubbing the cows, it can turn into a water fight. We make everything fun. When we once rented a barn that wasn't near our house, I'd cook a big meal and we'd have a picnic right in the barn. Once, when we had a picnic in the yard, the swing broke and fell on George and me - our sodas and chips went every which way and we laughed and laughed. That's what it's all about."

Even having to work three hundred and sixty-five days a year doesn't dampen their love for life on the farm. They willingly take on the commitment to be able to have this life style. The Bradts are milking seventy two to seventy five cows at the present time. The majority of their herd is Holsteins, but Eddie prefers Ayrshires.

"I don't like Jerseys, because they have attitude problems," he says. "My dream is to one day own one hundred registered Ayrshires. They don't give as

---

2 The annual fair held every August at the Sunshine Fairgrounds in Cobleskill.

much milk as Holsteins but they compete better in the ring and are much better natured. Also, there are too many good registered Holsteins around already for me to compete. But I just really like Ayrshires."

His favorite, Taffy, competes well, cooperating completely once she and Eddie are in the ring even if she has acted up and been a "pain in the neck" before they left the farm. Taffy has won Grand Champion in the Junior Grade and a lot of first place ribbons. Eddie's goal for Taffy is for her to win the Open Grand Champion. He wants to have one of the best Ayrshire herds in the country. George doesn't really have a favorite cow, nor does Ed.

"I see them all as employees," Ed explains. "If I see they aren't producing, they have to go. That's how I look at them."

Each of the three men has his favorite place on the farm. Ed likes being in the fields best, while his son prefers to be in the barn with the cows. George also prefers the outside work, so he can look for wildlife.

"Really, you must do both," Ed says. "There is a lot of work to be done in the fields, but the cows come first. And you need to know your cows. They are what make us money so they need to be number one."

"If you are invited to a wedding and one of the cows gets sick or has a problem, you aren't going to that wedding," says Terry. "You need to know that can happen and be okay with it."

All four said they like being able to work outside except in the cold winter weather, when the pipes freeze. Another down side of the business is that you are told what you'll be paid for your milk.

"Only in farming are people told what they will get for their product," Ed complains. "You can scream and write letters and object, but nobody listens. The politicians are not on our side.

"I am concerned about where the dairy business is going. When the milk prices are low the whole community suffers. The ripple effect is huge - restaurants, food stores, parts stores, seed, equipment, mechanics - everybody loses business. In order to try to make more money when the milk prices are low, we now milk the cows three times a day instead of the usual two. It's more tiring for us, but it increases milk production by 1300 pounds per pick-up, which is every other day. It's good for the cows, too, as long as they have enough of the right quality food - they do eat more when they are milked more often."

Schoharie County loses about thirty to forty farms every year. The Bradts are concerned that no matter how much the farmers try to get the politicians to understand the impact of losing area farms, they just don't listen. Ed frowns before going on.

"I think that maybe they don't understand because few of the elected officials were farm kids. They don't understand how much we need farms, and how much they add to the economy of the area. And seeing farms sit empty - especially this one - just kills people, especially the older folks who grew up around here and used to play here, or come to dinner parties and other social events. If this farm ever sits empty or starts to get run down, everyone is going to feel bad."

No one more than the Bradts. Their dream of someday owning this farm has ended. Unable to compete with developers, they cannot afford the going price to keep the place they have rented and farmed so lovingly for the past four years. They are packing up and preparing to move not only themselves, but the cows, too.

Ed stands up and reaches for his cap.

"They can knock me down, but they can't keep me down."

It is time to get ready for the next milking.

• • •

UPDATE
JANUARY 2015

In June of 2009, the Bradts and their herd moved to Rosenberg Road in Sharon Springs. Unable to find a house and barn on the same property, they are now renting a barn over a mile from the house, which makes keeping an eye on the cows and all the necessary chores that much more difficult. Things have not gone well since the move. Terry explains.

"After the move we lost some of our best cows, because they had a difficult time adjusting to a new style barn. And renting land from absentee landowners from the city makes for poor hay, because the owners won't let us spread manure, and without that fertilizer the hay is of poor quality. Cows can't produce a lot of milk unless they have the right food, and we can't make good hay without putting manure on the pastures. Also, our gas bill is very high because of all the driving between the house and the barn and out to the rented hay fields, which are spread so far apart. We're not getting ahead like we were on the other farm. We decided one of us had to get an outside job, and neither Ed nor Eddie wanted to be off the farm, so I am now working full-time for SUNY Cobleskill in the bookstore. I'm still doing ceramics and Mary Kay, too. So we are still plugging along, determined to make it work."

George decided that farming was not for him and he left the farm to go back to mechanics. He is now working for his uncle in Middleburg and living with his parents in Greenville. He hopes someday to work for the railroad.

"He's still always here if we need him," Terry says.

Eddie took an internship through SUNY, which led to a full-time job as an artificial inseminator in Greenwich, NY, but "I missed the farm. I wanted to be with my cows." So after a year, he came back to farm with his mom and dad. He also graduated from SUNY Cobleskill.

His Taffy did win a Grand Champion at the Sunshine Fair, although he did not get to take her to the State Fair. She died in February of 2014.

"She was a good old cow," he says.

PHOTO PROVIDED BY THE BRADT FAMILY

AT LEFT: **RAYMOND BRIGGS**
PHOTO PROVIDED BY KEVIN BERNER
BELOW: PHOTO PROVIDED BY
BRIGGS FAMILY

SITE OF HOME
LIEUT. PETER YOUNG
THE PATRIOTS OF NEW RHINEBECK
MADE THE HOUSE OF PETER YOUNG
AND WIFE THEIR RENDEZVOUS DURING
THE REVOLUTIONARY WAR

RAYMOND BRIGGS

Carlisle

INTERVIEWED ON MAY 30, 2008

While traveling along Route 20 and approaching the border of the Town of Carlisle, it is easy to miss the sign that reads: "Entering Carlisle, the home of Ray Briggs." A local student of history, however, might just be on the lookout for more of the same, for the Briggs family have not one, but three historical markers in the area. Their family is part of the heritage and history of Schoharie County.

A second sign on Rock District Road, is an official New York State Historical Roadside Marker, which reveals the site of a home where Patriots would often meet during the Revolutionary War. This colonial home belonged to Lieutenant Peter Young (originally Jung), a direct ancestor of Raymond Briggs. Many of Ray's ancestors fought during the Revolutionary War and also in the War of 1812.

"I am proud to be a member of the Sons of the American Revolution," Ray says.

Further along the same road, the third marker points to The Rock House, a cave where Peter Young and his wife once hid from Native American Indians.

It is here, amidst his own family's history in the Town of Carlisle, that Raymond Briggs, son of George and Eva Dristle Briggs, was born in the year 1924. His mother's German Palatine[3] ancestors settled on this land in 1760, making this two hundred and fifty four continuous years of one family on the these acres. Ray is very proud of his heritage and even prouder to be the seventh generation farming the land of his ancestors.

"Farmers worked harder in the early days. In winter, there was ice to cut and haul to the sunken icehouse, lowered by a horse-drawn elevator, and then covered with sawdust. We cut and split all our own firewood by hand, using horse and sleigh to bring it to the shed, where it was stacked by hand. We cut our own lumber and brought it to the sawmill. Splitting fence posts, making maple syrup,

---

3 The Palatines were a group of German Protestants from the west bank of the Rhine River who in the 1600s suffered persecution by Louis XIV. Aided by England's Queen Ann, many were able to escape, and between 1708 and 1710 more than 2000 arrived in New York, many of them settling along the Mohawk River.

shearing sheep, butchering hogs and sheep, everything was done by hand - we hand milked by lantern-light and delivered the milk cans to the Dairylee plant in Cobleskill, where the Dairy Deli is now. When electricity came in between 1936 and 1938, that sure made everything easier."

Ray has a long list of good memories of his youth. He attended the one-room Rock District Schoolhouse, which is only a short walk down the road from the farm.

"Now it's a home; you can't even tell it was a school. When I was in school, those were the golden years." His memories of going to Rock District and Cobleskill High School are just a small part of what he considers a fun and happy life.

"All the boys had fun ice skating and tobogganing in the winter and swimming and fishing in our pond or in Schuyler Lake in the summer. And there was lots of square dancing, which was loads of fun. We showed cattle at the Schoharie County Fair; I loved the excitement of the show ring. I had a Grand Champion at Future Farmers of America, 4H, and at both local and state fairs. The Briggs farm had outstanding sires - I had my own herd sire - and we raised our own calves and saw our herd make progress."

After graduating from high school, Ray was encouraged to go to college at Cornell by his friend Roger Barber.

"My first month there I was one homesick farm boy! I had never been away from home before. But I graduated from the College of Agriculture at Cornell in 1951 and am an honorary Ag Fraternity member. After graduating, I took a job with the National Holstein Association as a Holstein Classifier, which took me all over the country."

For many years, Ray served as a cattle judge in state and county fairs in New York and other states as well. In 1952 Ray married Vicki Dubonnet from Johnstown, NY, a descendent of an old wine family in France.

"She was a city girl, but she loved me so much she was willing to try the farm life. She did learn to like it. I have four sons, two daughters, ten grandsons, two granddaughters, and four great grandsons. They are all members of or eligible for the Sons/Daughters of the American Revolution.

"A few days after we returned from our honeymoon in Canada, three heifers had to be led from the barn to the day pasture. I gave the smallest one to Vicky to lead, while my dad and I led the others. Vicky's heifer was nervous and became very unruly and tried to run away, causing Vicky to fall. When I went running over to help her up, she looked up at me and said, 'Ray, take me home to Johnstown.' I thought I'd lost her. But she stuck it out and became a good farm wife and mother.

"I always had great respect for the farm wives. They worked really hard. They had big gardens, and canned and pickled everything, even meat. When there was a big blizzard the ladies had enough preserved in their cellars to survive no matter how long the storm and its effects lasted. And they had big families, too, did all the laundry by hand and fixed all the meals. They worked very hard and were very self-sufficient.

"My father was still running the farm when Vicki and I got married, so I took a job at Cobleskill High School as the Ag teacher. At the time this was the biggest Ag department in NYS. I taught there for twenty three years, and tried to instill a respect for farming, wildlife, and environment, in my students."

After his father retired, Ray ran the farm with his brother. For twelve years he was also an auctioneer at Walter Wagner Farm Sales in Central Bridge.

"That was a very good income for a farmer in those days."

Ray's beloved Vicki died in May of 2003. His son, Buddy, has the farm now but Mr. Briggs is still very active.

"The farm has been my life and I have loved everything about it - the animals, the farm environment, and the freedom of country life. I love observing nature, especially bird-watching."

Ray has been very influential in saving the New York State Bluebird population, and is founder and past president of the Schoharie County Chapter of the Bluebird Society as well as a member of the NYS Chapter. He received an award from Governor Pataki and is honored that his initials are on all the state Bluebird license plates. His own plate, #101, was specially made for him. In March of 2002 he was featured in an article in the magazine, Highlights for Children, which shows a picture of him receiving that award from the governor.

Ray is both a history buff and a writer - he wrote 50 of the 60 chapters for the Carlisle 200th Anniversary book and has done a lot of work on genealogies. He has lectured on the War of 1812 all around the area. Ever the historian, he gave me a short lesson on the history of farming:

"There was the Industrial Revolution and then there was the Agricultural Revolution - farmers first worked with their hands, horses, and pitchforks; then the tractor came along which was an outstanding labor-saver. Then came the square baler, the greatest invention since the wheel. After that came mechanized corn choppers and reapers and bundlers. Farms that formerly could only handle twelve to twenty cows could now easily milk thirty. Today, the average is eighty to one hundred, and even much bigger operations. We have bigger machines and can do more work. The small successful farmers could buy more land, incorporate other small farms and still keep up with the work. The most cows we ever milked on our farm was seventy."

Ray stopped and took a big breath.

"Farming in Schoharie County has diversified over the years. Hops were a booming cash crop until Blue Mold decimated it in the years 1912-1915. After that, dairy farming grew for several decades. Now there are many hobby farms. People move here from the city and buy a small farm to grow organic vegetables and fruit or have sheep, alpacas, or turkeys, whatever. That's a good thing. It keeps farming alive in the county.

"I have no regrets, but I do feel bad that we've let so many of the Old Dutch and Early Palatine barns go. There's not many left. Some were neglected and left to fall down. Others were purchased, dismantled, and rebuilt by people in Connecticut, Pennsylvania, New Jersey and even Texas and Colorado. It's a shame we're losing them. There's not enough preserved right here in the county. Just last year six old barns were dismantled and taken to other states.

"On our farm we did build a new barn in 1974. I still remember my boys leading the cows from the old barn. It was quite an event. The cows didn't mind the move because they had bigger, better, more comfortable stalls.

"I've been through it all," Ray repeats several times. In addition to being a farmer and historian, he has a strong feeling for the conservation of the environment and its wildlife. He is a history buff with an eye to the future, working for the conservation of endangered species not only of birds but also of plants. Just this past spring he took a trip up Bouck's Zourie[4] looking for a particular pink flowering shrub, which was decimated in the horse and buggy days by Sunday afternoon picnickers.

"I can't remember the name of it right now. I did find one surviving shrub, but I did not have time to look around for any others." He hopes to go back up the mountain soon and have another look.

When asked about his greatest accomplishments, Ray summarized his long and impressive list. They include: continuing the farm heritage and instilling the importance of farming and conservation in his children and grandchildren; graduating from Cornell; teaching; researching history and genealogy; working with the Bluebird Society.

---

4 Bouck's Zourie is the name of the tallest hill in the area, which can be seen from the Briggs farm.

"Best of all, I'm still able to be active on the farm after retiring and turning the farm over to my son. I still fix fence and cut brush and do whatever I can. Maybe working hard all my life made me strong enough to still fix fences and cut brush at 84 years old."

Asked if he remembers any more specific interesting events, Ray thinks for a time. Finally, he answers:

"When you've lived 84 years there've been lots of incidents, but I can't think of a one right now. I remember lots of things when I'm out fixing fence. That's when things come to me."

Ray Briggs is a remarkable man. How many people do you know that have three historical markers to their credit?

And like all farmers everywhere, he has remarkable grit.

* * *

UPDATE

Ray Briggs passed away in August of 2011. His son is now renting out the "new" dairy barn to Dan Ward. The old barn, with its wooden silo, is in a state of disrepair.

RAY RECEIVING AWARD FROM GOVERNOR PATAKI
PHOTO PROVIDED BY KEVIN BERNER, OF THE BLUEBIRD SOCIETY

THE IDEAL COW

I long for the cow of modern make,
That milks five days for leisure's sake,
That sleeps on Saturday, rests on Sunday,
To start afresh again on Monday.

Oh, for a herd beyond suggestion
Of staggers, bloat, or indigestion,
That never bothers to excite us,
With chills or fever or mastitis.

I sigh for a new and better breed
That takes less grooming and less feed;
That has the reason, wit and wisdom
To use a seat and flushing system.

I pray each weekend, long and clear,
Less work to do from year to year,
And cows that reach production peak,
All in a five-day working week.

Oh, why don't the scientific bobs,
Firmly entrenched in their cush jobs,
Show these ignorant breeders how
To propagate a five-day cow.

PRINTED WITH PERMISSION
BY BROWNE'S DAIRY PRODUCERS NEWSLETTER, PERTH, AUSTRALIA

DONNA AND JAY BURR

JOHN AND DONNA BURR

Sharon Springs

INTERVIEWED ON JANUARY 14, 2010

It was the love of cows and country life that made the Burrs decide to be farmers. John, usually known as Jay, smiles, as he tells why:

"I like the idea of being my own boss and having no one to answer to but myself. I like the feeling of accomplishment I get by having brought in all the hay, having the silo full, raising calves successfully, and solving a cow's problems."

Jay Burr was born in Brookfield, Connecticut on October 14, 1934. He grew up on a dairy farm at a time when the cows were milked by hand, the chores done without the benefit of modern equipment, and the teenagers themselves ran the 4-H clubs. It was at the 4H club that he and Donna met. Donna was born in Bridgeport, Connecticut on August 12, 1939. Her father was a tree surgeon, not a farmer, but had a cow, turkeys, chickens, and grew acres of asparagus. Donna and her brother would pick asparagus in the early mornings and after school they would put one and two-pound bunches in a red wagon and sell them up and down the road.

After two years of college, Jay left the family farm and worked for the Farm Bureau in Hartford. He soon tired of that.

"I couldn't stand road work and was ready to get back to farming." In 1961 he and Donna, then a schoolteacher, decided to get married, and then for one year they rented a farm in Pine Plains, NY. Then they returned to Connecticut to rent a farm in Kent for another six years.

By 1969 the couple was ready to buy their own farm. They worked hard that summer to manage to take two days to farm hunt, deciding to look in New York thinking it would be cheaper than Connecticut. They had expected to stay east of the Hudson River, but the real estate agent kept directing them to the Fonda/Fultonville area. The first day of their search was taken up by travel and talking to the agent.

"The Farm Credit Office closed at 4:30 but a man named Roy was still there when we got back. The next day he took us all around Montgomery and Schoharie Counties but we still didn't find what we wanted. Finally, he drove us to Engelville Road in Sharon Springs, stopping at the top of a hill. He pointed to a farm which sat on a neighboring hill, and told us it was not for sale but would soon be going into foreclosure." Jay smiles again. "We liked what we saw."

It was a picturesque place, with steep banks slanting down either side of the narrow driveway that leads up to the farm. A stream works its way through an adjoining field and runs lazily through a large culvert pipe under the driveway. The house and barns sit on the hill above the stream. Jay goes on:

"When the foreclosure did happen we bought it, borrowing $60,000 to do it. We were young and brash enough to think we could buy a farm, and we've lasted here for thirty nine years."

One winter day they loaded the kids - John, Jean, and Julie - into the car and drove over to New York to show them their new home. It was four days after a huge blizzard and they could only get as far as the hill where Jay and Donna had first seen the place the past summer. It was impossible to get any closer, so they went back to Connecticut and prepared to move.

Moving day finally came on April 30,1970. Jay came first, hauling thirty cows with the help of a friend, a cattle hauler. "It's nice to have good friends," he says. Donna followed the next day, leaving the children in Connecticut with a friend who brought them later.

"The car was too full to fit the kids in. My job was to get the house in order. My dad brought the furniture and everything got scratched because he packed it all in so tight. Julie was only one and a half, and when I brought her into her room and she saw her crib - still in pieces on the floor - her thumb went right in her mouth. She knew she was home."

After coming to Schoharie County the Burrs had another son, Steven.

"Being on the farm is the best way to bring up kids," Donna says. "They develop a good work ethic. They really learn how to do many kinds of things. From learning on our machinery, my boys could fix anything. When John decided to go to Nashville Auto Diesel School he tested out of four courses."

The Burrs worked the farm with mostly just the family. One son loved the fieldwork but didn't so much care for the cows. But their daughter, Jeanie, loved working with the cows and after graduating from Cornell took over the farm when her parents retired. She ran the farm from 1993 to 2000.

"She really knows cows," her father says.

Jay and Donna have a real passion for caring for the cows. Like many farmers, they often did their own veterinarian work, bringing cows through a variety of health crises. They each had their favorite cows. Jay talked about one who was blind "on and off" (they never discovered why) and how they accommodated her by always having her in the same place and giving her the extra time needed to 'feel' her way. Donna's favorite was a calf they raised that they named Red because of her coloring.

"Red knew her name and always responded when I called her. When the herd got too far from the barn I'd call for Red, who would come back when she heard her name and the other cows would follow her."

Though they enjoyed farming, it was sometimes tough, especially when the weather didn't cooperate.

"Once hay is down, it must be brought in," Jay explains. "Once when we had a thousand bales to be brought in and the weather report called for rain the next day, the two of us struggled to get the job done that evening. When it's going to rain, there's no dew, so we worked past dark." Jay drove and Donna struggled to catch the bales in the dark. But they got the job done.

The Burrs employed temporary hired help from time to time when it was need-ed. Though there have been many good workers, some did pose a problem. For instance, one young man stole some tools and a bankbook. But most were honest and helpful. One even pulled a pig out of a neighbor's barn while it was on fire.

Coming home from town one icy winter day, they saw the manure wagon crosswise on the narrow part of the driveway, the tractor hanging over the side of the steep bank. The distraught hired man didn't know what to do.

"Jay is a great problem-solver," Donna says. "He solved this problem by sepa-rating the tractor from the wagon with a sledgehammer and rode the tractor down the bank and across the ice back to the barn."

"Whatever the conditions are, farming is never easy," Jay says. "You have to work seven days a week and it takes a lot of planning to take time off. But farm-ers make good neighbors who help even if you need them at the last minute. That helps a lot, too."

Jay and Donna are those good neighbors to others. They are compassionate people. In August of 2005, after Hurricane Katrina, they traveled to Biloxi, Mis-sissippi with a group from their church, to help clean up after the terrible floods. When asked what they did there, Donna answers:

"Whatever needed doing to get things going for the people again."

More recently, there has been destructive flooding closer to home, in Fort Plain, and the Burrs are now making plans to assist with cleanup efforts there.

They are also politically minded and put their shoulders to the wheel to work for the good of farmers in the state. Jay served on the Farm Bureau Board for eight-years, Donna is past-president of the Farm Bureau. Together they have tried to get New York into the Northeast Dairy Compact, but without success, and are currently working on Farm Labor laws. They make frequent trips to Albany to meet with legislators, but are finding it very frustrating because, as Donna puts it,

"There is a big disconnect between farmers and non-farmers."

In such a meeting just recently, someone asked why farms couldn't work like factories do, with a forty-hour week and time and a half or double time pay for overtime.

"Stupid question!" Donna exclaimed. "Those people don't get that we work with live animals that need you seven days a week, and machines that break down and get you off schedule."

The Burrs continue to lobby for farmers' rights and prosperity, though their arguments often fall on deaf ears. In a recent letter to the editor of a newspaper, Donna wrote a letter that is concise, well thought out, smart, to the point, and occasionally sassy.

"Agriculture is a necessity," Donna states confidently. "But it gets harder and harder to make a living. I looked back through our records and brought the statistics to Albany to show them that we were paid less for our milk in 1997 than we were in 1980. But no one listens."

"We must have been on Easy Street back then and we didn't even know it!" says Jay, laughing.

"We moved to New York because we thought it would be cheaper than Connecticut," Donna says. "But that turned out not to be true. The first thing we noticed was that our medical insurance rates doubled. From then on we went without health insurance except for major/medical, choosing to pay for doctors' visits out of pocket. We paid as we went and ended up way ahead."

After their daughter, Jean, left the farm they tried renting it out to graduates of SUNY Cobleskill, but that never worked out. Jay got his truckers' license and for twelve years drove all over the Northeast hauling milk. One day, crossing the Tappan Zee Bridge in stop and go traffic, he hit the brakes and the hatch flew open. He lost some milk, and another driver yelled, "Got milk?" Jay smiles at the memory.

In 2008 they decided to sell the farm. They now live down the driveway in a smaller house on a parcel of land that had been broken off the original farm, and built from timbers of an old barn that had been torn down. After the closing, the new owners told the Burrs that they planned to knock down the farmhouse and put up a new, smaller one. Donna was heartbroken. The day the bulldozers started on the job Jay had to put her in the car and take her out for the day.

"It was very upsetting," she murmurs softly.

Jay and Donna have two grandchildren and two step-grandchildren. But they also have had more than their share of heartbreak. They have lost both sons in accidents, one involving a four-wheeler, the other in a car crash. But they live with their sorrow and go on trying to do what they do best, working to improve the lot of other farmers and helping neighbors.

"Our greatest accomplishment was not to go broke," Donna says. "There were once farms all up and down the road, now not so many. But we were able to hang in there.

Jay says he wants to be remembered for always smiling, for being a decent person, and for doing a decent job.

ABOVE: **VAUGHN AND PHYLLIS CREWELL**

BELOW: **PHYLLIS MILKING**
PHOTOS PROVIDED BY THE CREWELL FAMILY

VAUGHN AND PHYLLIS CREWELL AND SON, MIKE

Schoharie

INTERVIEWED JANUARY 27, 2010

Overlooking the valley between the villages of Cobleskill and Middleburgh, the Crewell farm sits nestled in the gently rolling hills west of the Schoharie Creek.

Vaughn Crewell was born on a dairy farm in Cherry Valley on April 24,1936. As a young man he worked for a cauliflower farmer and also as a truck driver, but then decided to return to milking cows. In 1961 he bought thirty head of cattle and rented a farm in Cherry Valley and six years later he was ready to own his own farm. On October 5, 1967, he came to Schoharie and bought twenty acres, complete with a barn and some old outbuildings, from Claudette Wood, a relative of the original owner, Jacob Borst. Years before, Jacob had split up his original farm so that each of his children would have a piece. Vaughn gradually bought back many of these smaller farms and now owns forty-six acres and rents even more of the former Borst property.

Phyllis Armlin, a descendant of the Palatines of Pennsylvania, grew up on a small farm in nearby Middleburgh. When she and Vaughn met she was a single mother of four - Margie, Ernie, Lisa, and Mark - and working as a nurse's aide in Bassett Hospital. After she and Vaughn married in1969 their family began to grow and together they had three more children, Mike, Scott and Jessica.

"Having seven children follows the tradition of our families," Phyllis says. "My mother came from a family of seven and then she had seven children; Vaughn also came from a family of seven." They have nine grandchildren now, and are very proud of them. The Crewells consider the farm to be a great place to raise children.

"Kids on a farm learn to stand on their own two feet," says Vaughn. "Together we built this whole place. The house is the only original structure and we did a lot of work on that, too." Family is important on the farm and the Crewells reminisce about how extended family members have helped them with many different projects, including re-siding the big barn and finding old beams to replace the original ones that were no longer strong and useable.

In addition to taking care of the family and the cows, Phyllis also loves to garden, and raises her own vegetables.

"I needed a new rototiller last year, but it was a tough year and I was down to my last eleven dollars. So my garden grew up and …….." Rather than finish the sentence, she shook her head and threw her arms up in defeat.

Vaughn and Phyllis agree that farming is a lot of hard work. At one time - when they were younger - they milked one hundred and fifty cows. Though their main barn only held fifty-seven cows, by putting in a pipeline and a bulk tank they were able to milk them all in three stages. They kept the best producing cows in one barn, the mediocre milkers in another, the dry cows in a third. In recent years, however, they have downsized their herd.

Vaughn says that farmers need to have a wide variety of skills:

"A farmer has to be a veterinarian and mechanic, too, and needs to be able to find a solution to any kind of problem. I like working with the animals and the land, but I'm also a hay dealer."

"Farmers also have to have a sense of what has priority," Phyllis adds.

Though the children worked right along with their parents when they were growing up, at times farmers also need even more hands. But like many other farmers, Vaughn says they often had a hard time finding good hired help.

Mike agrees. "We've never had many other animals besides cows, but we've sure hired a lot of jackasses!"

In addition to some other concerns with outside help, Vaughn's biggest complaint was that some hired men tended to underfeed the cattle.

"That really irked me. I had to tell them over and over to let the cows eat their fill before milking. The best hand we ever had was good because he really loved the cows and treated them right. We have often had veterinarian students that were assigned to our farm for ten-week periods. Some were girls, and they were generally the best workers."

With a touch of pride Phyllis adds, "Vaughn really knows cows. He knows what he's doing and he likes the cows. We both do."

Phyllis herself has a favorite cow that she calls Knobby. When Knobby was a calf, she wrapped her neck chain around one leg, and though no bones were broken, the ankle was severely injured. Phyllis put on hot packs, wrapped the leg, put salve on it, and gave the calf aspirin in a handful of grain. Then she did therapy every day, and Knobby is now completely healed and has had two fine calves of her own.

Though both Marge and Ernie liked the cows, none of the children chose to continue in farming except Mike, who didn't care for milking cows but did enjoy working in the fields.

"Why work for nothing? I got off the dairy as fast as I could to work just the crops."

Mike is also a good mechanic and has done all the machinery repairs since he was a teenager, which has been a big financial help to his parents.

"It's expensive to hire outside mechanics," Vaughn says.

"Mike can do just about anything," his mother brags.

Always a good worker, when he took over the crops Mike became an even bigger asset to his parents. He and his wife, Amy (Prokop) Crewell, bought the farm next door and now are in the feed business. Amy owns and runs the adjoining ice cream stand.

Vaughn says he has no regrets about having chosen the farm life, but adds, "I don't like what we get paid for what we do and I don't think we'll ever see a change. Even if you love what you're doing, it sure would be nice to always be able to pay your bills.

"Right now is a very stressful time. This week milk is up and we get paid $17 a hundred pound, but it will go right back down. But when it costs about $16 to produce a hundred pounds of milk, how do you make it? We've had both good times and bad.

"One of the bad times was when we lost our barn to a fire on May 30, 1992. We lost thirty heifers. Fifteen burnt in the fire and the rest had to be put down 'cause their insides were too badly damaged." Sadly, a former childhood friend of Mike's had set the fire. He was convicted and served seven years in jail for arson.

At the present time, Vaughn and Phyllis are milking only thirty-three cows. Vaughn has had a few surgeries, including a hip replacement, has many aches and pains and can't do the work anymore. For two or three years Phyllis has been on medication for a problem in her spine that decreased the blood flow to her brain. She also is always in pain.

A year or so ago they sat down and talked about retirement. Going over their finances they felt they would be okay living on their Social Security and Phyllis's small pension, but then the price of milk fell and they could no longer afford retirement.

"The cows won't bring in enough money to make it worth selling them," Phyllis explains.

Asked what was their greatest accomplishment, Vaughn answered without hesitation:

"Having raised seven kids, all of whom are smart and very successful."

They both would like to be remembered for being dedicated to farming.

• • •

UPDATE

FEBRUARY 2015

Two years after this interview Vaughn sold the cows, as Phyllis had developed Alzheimer's, and he could no longer do the work alone. He then spent his days caring for her until she died on July 6, 2014.

"It's been very hard," Vaughn says. "We were married for forty-seven years."

Vaughn says that having two of his sons and a daughter living nearby has helped him cope.

"Plus, I have a new one-year old grandson, McCormick, Mike and Amy's son, and I have him here with me almost every day. That helps a lot."

Mike has taken over the land and continues to grow crops. In the summer Vaughn stays busy helping bring in the corn and the hay.

VAUGHN WITH HIS GRANDSON, MCCORMICK CREWELL

PHOTO PROVIDED BY THE CREWELL FAMILY

PATTI AND TIM EVERETT

TIM AND PATTI EVERETT
STONEHOUSE FARM

Sharon Springs
INTERVIEWED ON APRIL 19, 2013

Drive along Lynk Road, which curves through wide flat fields spreading out on either side, and you will come to the farm of Tim and Patti Everett. Coming up a small rise, the grand stone house comes into view on the left, a substantial dairy barn sitting beyond. Directly across the road from the house sits their rustic sap house. Welcome to Stonehouse Farm.

Farming has always been what Tim Everett knew best. Born in 1959, Tim grew up on this very farm, which had been purchased by his parents in 1964.

"After high school I knew I wanted to continue to farm, so I went to work on my uncle's farm, since my two older brothers were already working the home farm and there wasn't room for all three of us. But my uncle farmed differently than the way I grew up and I wanted to do things my own way. I like being my own boss and doing things for myself. I have always liked trying to do different things; I get bored easy, so sometimes I start something and then move on to something else, and then to another."

Patti, also born in 1959, grew up on the other side of Sharon Springs on a dairy farm owned and run by her parents, Harold and Barbara Hall. Thinking she might want to be a teacher, she decided to go to college.

"But I found out that sitting at a desk inside was not for me. And I sure didn't want to drive a long way and sit in an office somewhere for my job. I like that on the farm every day is different. I love the fresh air. As long as the cows are milked I can step outside anytime and do whatever I want to do. So after one semester, I realized that what I really wanted was a farm family."

Tim and Patti married in 1978, and have two daughters, Randi and Amy. After seven years on his uncle's farm, Tim's parents asked if he and Patti would take over the home farm.

"We were doing well where we were," says Patti. "And the girls, who were little then, didn't want to move to another place. They both cried and said 'We don't want to live in that house!' But after talking about it for awhile we decided to do it."

"It was not an easy move," Tim adds. "But I knew the land on my home farm, and it was better land. And it was a newer barn. My dad wanted us to keep both farms, but I didn't want to have to hire any help and deal with any employees. Like I said, I like to do things my own way, and you can't manage that if you have someone else doing the work on the other farm. If we had today's equipment, we could have run both farms ourselves, but back then, most of the work was done by our own hands. So in November of 1986 we bought out my folks and took the farm over."

"The place wasn't in good shape when we took over and it took three years to clean everything up," says Patti.

Tim and Patti have had their share of challenges on the farm. Eight years ago their home burned completely due to a chimney fire.

"We lost everything except the stone walls," Patti says sadly. "Since Tim's parents were away for the winter, we lived in their house 'til they came home in the spring, then moved into the sap house while we rebuilt the house. It took us months.

"Another big challenge was continuing to keep the farm going when the girls were teenagers. They saw what non-farming kids were doing and wanted the same things. We made sure they took their music lessons and did their sports, but it was hard to fit it all in sometimes. They learned they couldn't do everything and were often disappointed, but they grew up fine anyway and got over it. They learned not to complain, because they found out it wouldn't do any good. They learned to have a sense of responsibility - and being in 4-H also helped with that. Our daughters are real workers. They still help us out a lot. Every fall Tim and I go to Vermont to get maple supplies, even stay over in a motel once in awhile, and while we are away they do the milking for us. We went to Florida once and to North Carolina to visit Tim's sister, but it is better to be home.

"Randi is now married to Tim Karona and they have a two-year-old daughter, Maya. Tim has also been a tremendous help. He does some of the fieldwork and with his help we have increased our cropland so we can sell more hay. That's good, because there is more money in selling hay than in selling milk, and we can control what we get paid for that."

"My son-in-law likes mechanics, so now he does a lot of the repair work, which is a great help," Tim says. "I don't like mechanics, but I had to do it for years and there was no sense whining about it. It's better to maintain your equipment well, take care of stuff first, so it doesn't break down. It is a lot easier to use a grease gun than a welding tool! Breakdowns cost money and time, so I take good care of my equipment; it is a necessity. We make things pay for themselves, and don't borrow money. We only buy new if the old is worn out and can't be repaired anymore.

"If I had today's knowledge when I was nineteen, I might have done something else. You lose money on the cows, yet they are what paid for this farm. So probably, I'd do it again. One thing is for sure: farmers are going to continue to struggle."

Patti proudly shares one of their biggest accomplishments:

"In 1995 we reached a high when our farm was listed as one of the top ten in the nation and received the Young Jersey Dairyman's Award. Our farm was on a national tour. People from all over the country and even other countries came to see it. We were even on the internet."

Aside from their pride in having one of the top herds, there have also been certain individual cows that were favorites.

"They really do have different personalities," Patti says. "Our first favorite belonged to Tim's uncle – she lived to be fourteen years old and always produced well. My nephew liked one called Daisy, and compared all and every other cow to her. Now, we have Barb, who is our highest appraised cow ever. In 2008 our daughter Amy, with the help of a friend, took Barb to a big show in Louisville, Kentucky - the Kentucky National. But once they got there, Barb wouldn't perform, wouldn't even go into the ring. Then, she got sick and wouldn't eat or drink. She was in bad shape and they needed to get her back home quickly. They were so afraid they were going to lose her that they stopped every hour to check on her and try to get her to eat or drink something, but she refused. She got worse and worse and they drove all night to get her home. Once she was back in our barn, she right away drank some water and ate some food and was fine. I guess she was homesick!"

Over the years, the Everetts have seen many changes in the dairy farming industry.

"Shipping milk through the co-ops is not always good," says Tim. "Years ago, when we were still on my uncle's farm, we got a call one evening that the North East Dairy Co-op, which we were then shipping to, was going bankrupt. The co-op owed us a month and a half worth of money and never paid us. Losing all that money made us leery of co-ops for a long while. That's when we started shipping our milk to Cappiolo's Cheese factory in Schenectady, which we did for the next ten years. Cappiolo treated us really well; they gave us a cash advance and picked up our milk every day. They liked our milk because we have all Jersey cows, whose milk has a high protein content. A good protein content is needed for good quality cheese, and our Jerseys produce milk with good high protein."

Tim explains why the term butterfat is no longer in vogue when discussing the quality of Jersey milk: "People have become very health conscious and avoid anything with high fat. But at the same time, they are looking for high quality protein in their food. Jersey cows fill that bill. The difference is purely psychological.

"Ultimately, we felt we had no choice but to go back to shipping to a co-op because that's how the industry is set up. We shipped to Dairylea for awhile, but they kept getting bigger and bigger and tacking on more fees and assessments, taking more and more of our milk check profits - up to 10%. So we recently left them and now we ship to the Cooperstown Cheese Company."

"Another thing that has changed is the demand of the public," Patti says. "Most milk is sold at the convenience store now. People want convenience; they don't want to go to the farm or even to the supermarket. And now you need a special permit to sell raw milk."

"The milk industry is all processor controlled now," Tim adds. "The farmers don't have much control. Low milk prices have driven the industry, and the government tells us what we can get paid. I don't think there's any future for the midsize farm. You have to milk a thousand cows to make any money at it, and then you have to deal with employees. The little farm can hang on and the big farms can make a living, but the mid-size has no future. We're seeing the results of the last couple of years - a larger portion for the processor, less for the farmer. We are a little farm - we're milking only 24 cows now. A mid-size farm milks around 300 cows. We were milking 48 cows up to two years ago, but we are getting older, and once the maple syrup business took off we downsized the herd."

"Mother Nature has thrown a lot at us, but I don't see it as being a challenge. I enjoy it all," Patti says. "I love being in the woods gathering the sap - it's peaceful. It's amazing how little most people know about how trees make sap and how we harvest it. Because it really is a crop."

"At first we made maple syrup the same way my grandparents and parents did it – in a big kettle outside," Tim says. "But a lot of the time the syrup was awful! They would just keep boiling and adding sap and it tasted burnt. I didn't like that syrup at all. One day I took $1000 without Patti knowing it and bought an evaporator. Patti was not happy when I brought that home! In 1994 we built the sap house, and started the pancake breakfasts and that business has really taken off. It helps make up for some of the money we lose on the cows."

The Pancakes in the Saphouse weekend breakfasts have become very popular. Both Tim and Patti - as well as their daughters and Patti's father - can usually be found here on those late winter mornings making pancakes and waffles, sausage and bacon. Folks sit at long tables, eating leisurely while visiting with friends and strangers alike. Others gather around the steamy evaporator, soaking in the warmth of the wood fire and watching the sap slowly boil down into delicious syrup.

"Last year at one of our breakfasts, a little boy asked us how we get the sap out of the tree after we cut it down," Patti says with a smile. "I had to explain how we tap the trees and collect the sap. We work really hard during maple season and someone recently asked me if we would take a month-long cruise after it was over. I said, 'Sure. On the tractor!' We've traveled a little, but I prefer to be home."

Tim and Patti talk fondly of having watched the girls grow up and are now reliving that same experience with Maya.

"It has always been special to go to Grandma and Grandpa's farm, to find new kittens, and jump in the hay and all that," Patti reminisces. "That's what it's all about. But it's still uncertain whether or not either of the girls will want to continue on the farm. They don't want that responsibility. Yet."

Tim adds that there's no use in complaining.

• • •

UPDATE

The Everett's happily report that they have a new grandson, Kaydem, who was born on December 3, 2014.

FRED AND EVIE GARBER

FRED AND EVIE GARBER

Cobleskill

INTERVIEWED ON MARCH 3, 2009

Fred Garber remembers when his grandfather and father worked the farm with horses. When one of the horses died they kept the other as a pet for Fred to ride, but the horse's back was too wide for Fred's little legs to straddle. Gradually, tractors took over from the horses, making the work easier but creating a big financial burden on the farmers - since all the equipment had been designed and manufactured to be pulled behind horses, everything had to eventually be replaced.

Farming was in Fred Garber's blood and it was all he ever wanted to do. He feels that the best part of farming was being his own boss.

"We could make our own decisions - for good for bad, they were *our* decisions." This is a recurring sentiment in my interviews with Schoharie County farmers.

Fred Garber and Evie Baker grew up on farms in Cobleskill and Carlisle, respectively. They both graduated from Cobleskill High School. They married in 1964, had three sons, Randy, Brian and Chad, and in 1967 bought his Dad's farm on Hallenbeck Road. Fred was the fourth generation to farm that land - his maternal great-grandfather had purchased the farm in 1917, which then passed on to his grandfather and then to his parents, Maynard and Beatrice, when Fred was four years old. Fred had the opportunity to go to college, but chose the farm life because he loved it and it just seemed natural to him. He says he learned enough in the high school Future Farmers of America Club and 4H to go right into it.

Evie grew up on a farm on State Route 145 just north of Cobleskill. Her parents, Mary and Howard Baker, bought the farm in 1948 at a tax sale, and Howard continued to work at General Electric in Schenectady until the farm was paid for. He also served as Supervisor for the Town of Carlisle for many years.

Evie and her three sisters did farm work from the time their father first put them on a tractor at age three. They would steer while Dad picked rocks, which are in abundance in this part of the country. Evie remembers a few minor mishaps while behind the wheel, such as tipping hay bales off the wagon, but no major problems despite small girls handling huge equipment.

In addition to the cows, Fred and Evie had pigs, chickens, turkeys, ducks, and sheep. The animals supplied them with eggs and meat, which came in handy when they needed to feed the family and money was tight. Summers, they would hire local boys to help with putting in hay and other chores. Evie says that if you worked with them and taught them what you expected, most kids didn't mind working.

Fred's favorite part of farming was in the barn milking the cows. When in his forties, his back gave out and he was no longer able to do it. After that, Evie milked the fifty or so cows by herself and he did the fieldwork, mostly driving tractor. When asked, he said he didn't have a favorite cow, but Evie reminds him of Denise, who would talk back to him when he spoke to her.

"Oh, yeah," he says, smiling. His eyes twinkled for a moment. "Some you do remember."

"There were some cows that had unique traits, like Beauty, who acted as though having her back scratched was like being tickled, and she'd wiggle around," Evie says. "I was really more interested in the kids than the cows, but I did doctor them a bit."

"That was the part of farming I didn't care for, when the cows got sick," Fred adds. "Not only did they stop producing milk, but I hated to see them suffer."

For Evie, the best part was family. "Staying home with the kids made me happy but also made the kids happy. We could always be together, whether working or playing, which many times was one and the same.

"When we would go out to fix fence, I'd take a picnic lunch, and maybe we'd go swimming. They loved to go swimming. After we built the pond in '82 they swam every day, and when our son Chad went into the Army he said that swimming was the best thing he'd ever learned. When he told me that he had to jump out of planes into water with a heavy pack on his back, I was *really* glad he had learned to swim.

"Fun on the farm sometimes came from unexpected sources. A friend had a pony to give away and the boys begged to have it. Fred and I talked about it for quite awhile and finally agreed. Right after our talk, we looked out the window and that very same pony just came walking up the road toward our farm. The pony had broken out of its fence and just came here on its own! I said, 'Well, I guess it's time to have that pony.' "

"A great place to raise kids, the farm," Fred says. "The boys had to take turns with doing outside activities like scouting or sports, because we couldn't be running them back and forth all the time. I told them they could continue as long as no bad things happened…if I heard that they did anything bad, they couldn't go any more. I told them it was a lot easier for me to stay home than run them around, so they knew they had to behave."

"The boys got into enough trouble around the farm," Evie says. "Inspired by cowboy stories, they played cowboy and decided to try to rope the cows. They lassoed one heifer around one leg and it scared her so bad they all had a terrible time getting ahold of her and taking the rope off."

"Oh, that was in the spring," Fred adds, shaking his head and laughing, "and there was so much mud! It all happened in the mud. What a mess."

"But the boys learned how to work," says Evie. "All farm kids, including myself, learn how to work. Today, kids have no idea where their food comes from, and they don't know how to work with their hands. It's all computers today."

Family is still important to Fred and Evie. As a way of keeping relatives in touch with each other, Evie says they are having twenty-eight extended family members in for breakfast next Sunday morning.

"I don't know where I'll put them all, but we'll figure it out. We always have."

Fred says that they didn't get to go places much, but the family was always there. Family, extended and otherwise, was very important, as was being a good neighbor.

"In the late '70s, a neighbor got hurt in the fall of the year, and all of us farmers got together and brought our own tractors and equipment, and in one weekend cut all his corn and got it in for the winter. Evie expected me home for lunch that day, but I didn't get home till late afternoon. The wife of the injured farmer had cooked a huge dinner for us all, including homemade pies. Best meal I've ever eaten.

"When another farmer's barn burned down after haying season was done, every other farmer in the area took over one truckload of hay to insure he'd have enough to feed his cows all winter. Nowadays, people don't even know their neighbors' names," Fred says sadly. "Like everything else, farming has changed. Today, farming is sometimes more about business than family. But some changes are good, I guess. In farming things change, so you have to change."

Evie remembers when gas prices went up and she decided to walk to their second barn, where they kept the dry stock, instead of driving the farm truck.

"I was better off walking, anyway. We would plan all our chores in town for one day a week and do everything in that one-day's outing to save on gas. There was no time or money to do more than that."

Fred goes on. "In hard times, my father would remind us that 'If you don't have anything you can't lose anything,' so we were always grateful for what we had. You have to take care of your own, 'cause the politicians sure don't care about the farmers. And they won't care until they can't get any food for themselves. Economically, farming has gotten harder and harder. There are more farmers working on farms owned by other people, now. And it cost more to get out than to get in these days because of the load of debt incurred. We've been lucky. We both understand why none of our sons wanted the responsibility of owning and running the farm when we retired. They were always here when we needed help and they enjoyed the work, just not the ownership."

The Garbers sold the cows in 2000. Several farmers wanted to buy the farm but they couldn't come up with the money, even though the Garbers were willing to lower the price to help them. Sad that their land was to be broken up and developed, they took the advice they were given to sell it in parts and retain ownership of half, and built their new house on part of the land they kept.

"When we finally sold the farm in 2005 it was done on a handshake," Fred says with a smile, "and left that way for six months until the final papers were signed. I was proud to be told by someone in the community that this was acceptable to the buyer because my name has always been honorable. I would like to be remembered as always having been fair and honest and having the respect of others, and for having left the land in better shape than it was. The new owners have a hobby farm now, with pigs, chickens and goats."

For eleven years after retirement, Fred worked at the fairgrounds for eight months of the year, doing maintenance.

"They wanted me to be in charge of the board, but I had run a business all my life and wanted no more responsibility. Now I want to do my work and go home."

In 2012 he retired from the fairgrounds, but "I need something to do, so this summer I will mow lawns at the golf course. I love to drive tractor, so I think that will be good."

Evie drives school bus and supervises on the various playgrounds of Cobleskill-Richmondville School District. She started out on the playground, but was soon approached to drive bus because of her experience driving large equipment on the farm.

"I don't have any trouble with the kids because of great advice I was once given: that I have a lot of stuff in my head and I can't put my head on the kid's shoulders. I liked that advice and always remembered it. So I do what I did on the farm with the hired boys…I teach them what is acceptable and stick to it."

L TO R, LAURIE, FRED, EVIE, BRIAN, LINDA HOLMES.
TOP: CHAD, RANDY
PHOTO PROVIDED BY GARBER FAMILY

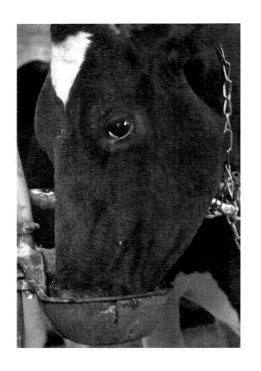

THE SILO

The silo stands empty, a lonely reminder of past, more fertile days.
It's slender shadow falls on once-furrowed fields
now fallow,
littered with the lint of milkweed and thistle.

No plentiful harvest fills its circular center,
Even mice have abandoned its round walls for more bountiful homes.
The path worn bare by cloven hoof and work boot,
now obliterated by weeds, is lost from view.

Sturdy barn walls, once alive with echoes of gentle lowing and
tinny sound of warm milk striking metal pail
have, like the farmer, returned to dust
marked only by a monument of stone on stone, laid by determined hands.

No mewing kittens or crowing fowl
call farmer and wife to daily chores;
no bawling cows, udders swollen with their offering,
no garden with its green abundance demands the hoe, then pays its bounty.

Now only stillness here.
A butterfly lands on Queen Anne's Lace.

BY

META WATTS

RUTH AND ART GRAULICH

ART AND RUTH GRAULICH
ARGUS ACRES

Sharon Springs

INTERVIEWED ON JANUARY 15, 2010

Argus Acres has had only three owners since it began in 1818, when a man by the name of John Hyney built a mill and a large brick house here. Mr. Hyney had a mule business on the barge canal and brought the bricks for the house down the canal from Fort Plain, then transported them to the site by ox cart. Mr. Hyney continued to own the farm until 1888 when he sold it to the Burgess-Stokes family. It stayed in that family until purchased in 1948 by Erhardt Graulich, father of present owner, Art Graulich.

A beloved aunt and uncle, who were dairy farmers in Dutchess County, instilled Art Graulich's love of dairy farming. Art was born on Long Island on October 11, 1929, and though his parents had a few chickens and a couple of sheep, it was through frequent visits to their Dutchess County relatives that he was introduced to working with cows. By the time he was fifteen he had developed such a passion for dairy farming that his mother and father moved from Long Island to upstate New York and bought a farm in Milford. Not knowing if their son would or could succeed, it was an act of love that has had wonderful results and justified their faith in him.

Art graduated from high school in 1947. The family then decided it was time to buy a larger farm, so on March 1, 1948, his father bought the two hundred acre farm in Argusville, where Art and his wife Ruth still reside today. Argusville in those days was a thriving community, with one hundred houses, three churches, and plank sidewalks. When Art was only nineteen, he bought another one hundred and seventy acres for $3000, money lent to him by his aunt.

Ruth Sturm, the daughter of a bricklayer brought up on a farm in Germany, was also born on Long Island. When she was eight years old, the family moved to a farm in Cobleskill because of the lack of construction work on Long Island. While in high school, Ruth worked for Nellie Gordon, a local attorney and tax preparer.

Art and Ruth married in 1953, and when his dad retired in 1963, Art and Ruth bought the farm for themselves.

The Graulichs have four daughters, Linda, Joan, Mary and Lisa, and one son, David. David and his family live in Ruth's childhood home on North Grand Street in Cobleskill, situated just outside of the edge of the village. The big red barn still sits grandly on the hill that slopes down from behind the big house and most days a horse or two can be seen grazing in the spacious pasture that spreads out beyond.

The original Graulich farm of two hundred acres has grown into an operation that encompasses fourteen hundred acres on four separate farms. Art is a good planner, and through estate planning the farm is now a Limited Liability Corporation called Argus Acres, with Art and his son David as co-owners. The family has grown along with the farm, and they now have fourteen grandchildren. They hope that some of the grandchildren will continue farming.

Reiterating a common theme, the Graulichs believe that raising children on a farm is a good way of life. Ruth has done some barn and field work, but Art says she has done the most important work of all: caring for the children and keeping the books. Their five children have become responsible, competent, common sense adults.

"The children had chores to do in a timely manner," Ruth says. "And they learned that you do things that you don't get paid for. Our kids can do almost anything. They all benefited from growing up on a farm. They are smart and have all done well - two valedictorians and one salutatorian. Joan attended Cornell and is now a veterinarian. Linda is Director of Education at St. Peters in Albany. Mary is a teacher and Lisa is a financial analyst."

Art and Ruth reminisce how their kids had to amuse themselves and make their own fun and entertainment.

"David built a cabin on the property when he was only twelve. In the process he cut down trees he wasn't supposed to, but I didn't punish him. We had a lot of trees," Art said with a smile. "David also built a houseboat, but that sank the minute he put it in the pond. We had to pull the thing out."

As a boy, David loved to ride his pony whenever he could, doing some tricks and telling about it in a paper he wrote for 4-H. Demonstrating one move, he wrote that he and the pony did a "fig-your-eight." Obviously very proud of their son, they laughed when they said that he had never been a good speller.

"No, he wasn't much of a speller," Art says, chuckling, " but he sure is a good man."

David had aspired to be a jockey and every year Art and Ruth would take him to Saratoga Raceway on opening day so he could walk around and talk to the jockeys. One of them took one look at David's hands and predicted that he would grow too big to be a jockey. The prediction came true and the day came that the pony was able to walk right out from under his long legs.

David attended Cornell for a short time but missed the farm life. Returning home, he went to SUNY Cobleskill, made good friends, learned people skills, and a lot about farming. One of David's successful projects is the beautiful Dutch barn he built that sits straight ahead as you come into the Graulich driveway. David also learned to fly and has airplanes, though his flying time is limited now that he is married and a father.

Art was Carlisle town supervisor for eighteen years, his wife working by his side as town bookkeeper. He has been on the SUNY Cobleskill College Counsel for fourteen years, but says he most likely will not be re-appointed because we now have a Democratic governor. He has been president of the Schoharie Dairy Cooperative since 1985.

Art talks while sitting next to a table piled with farming magazines, which he refers to from time to time. He is a man who keeps up with the latest and respects progressive farming.

When asked what he likes best about farming Art's answer is a familiar one: being your own boss.

"Of course," he adds, "any decision you make is your fault. But I love the animals and growing the crops. My favorite farm chore is the tractor work, which I continue to do."

At present, the Graulich farm milks three hundred and fifty cows, all registered and purebred Holsteins. Milking this many cows makes hired help a necessity, but David works right along with the hired men. Unlike many farms, Argus Acres has a history of keeping help for the long term. One man stayed twenty years and then his son took over for a number of years. Art also employs four Guatemalan men. He agrees with other farmers who complain that they can't find local men that are reliable and willing to do the work.

He laughs. "I could write a book on hired men."

Art regrets that the small farmer can no longer make it. When he started out there were twenty-six dairy farms between his place and Cobleskill. Now there is only one. To anyone starting out today he says, "Marry into it and don't quit your day job!"

"This past year has been one of the worst ever," Ruth says. "If we didn't have savings we'd be sunk."

"Or borrowing up to the hilt," Art adds. "I predict that the farmers who stay in business will be the ones who milk more than one hundred cows, a big increase from the twenty cows my dad began with. I have friends who have even bigger operations, milking anywhere from seven hundred to thirty five hundred head. Hired help and family working together is the only way it can be done."

Another change from the smaller farm is that cows are no longer given names; they now get numbers on the bigger farms. Art says he may have had a few favorites, but couldn't remember any specifically. However, he mentioned #81 a few times, making the interviewer wonder if that one hadn't been a favorite.

"With over six hundred head there are too many to name," he says with a hint of regret in his voice. But Ruth remembers one particular cow.

"She was just wild. She would chase the kids right under the fence."

Despite stories about the dangers of farm life and of injuries on farms, the Graulich children were never hurt. Ruth attributes it to their commonsense, a sense of safety, and lots of luck. Art holds up a hand, showing one finger missing its tip.

"Lost this part of my finger while building the barn," he said. "Cut it off with an electric saw. My only real injury."

This leads into another story about a neighboring farmer who generally worked alone. Their adjoining farms were separated by a woodlot, and they happened to see the man try to start one tractor by pulling it with the other. The towed tractor ran over the tractor in front and pinned the man up against the steering wheel.

"We got him out," Art says matter-of-factly. "He was okay, only had a bruised chest."

Asked what is their least favorite part of farming the Graulichs lodged a familiar complaint: no say on what you get paid for your product. This question started a long and familiar discussion about the changes in farming and how the government regulations make it so difficult. Art also tells how cows used to last longer.

"Now they push them too hard to increase production." Art looks down and shakes his head before continuing. "In 1948 the average cow could produce about eight thousand pounds of milk per year. Now, through genetics and feed, it has been pushed up to twenty five thousand pounds! It's too hard on the cows."

Ruth mentions another familiar complaint: farming is very confining for some people. However, Ruth says she never minded because she didn't really need or want time away.

"I'm a hermit, really," she adds in her quiet, serene way. "I like just being home."

One difficulty on Argus Acres is the wetland surrounding the barns and the house. It was because of the wet fields that Art started to buy other land. In early years they considered selling and moving again, but by then they had invested too much in both the house and the barns. Since wetland decreases the alfalfa crop, which is necessary to feed the cows, they needed land that was higher and dries out faster in the spring. So they bought two farms on higher ground along Route 20. Even so, the land on each farm varies, making it necessary to use different techniques for improving production.

"Some fields can be dry and others wet even in the same weather conditions," Art says. "Each year, the weather determines which fields we work first, and which crop we plant in each field. That way I can always cope with the weather."

The conversation turned back to how farming has changed in the years since they started. In 1950 artificial breeders started to take over the job that had previously been left to the bulls. Art has not had a bull on the property since 1980. Despite his many progressive views, he expressed concern over 'messing with nature' as in cloning of animals and hybrid crops.

"All the corn is hybrid now….after you harvest it, the seed will not grow again next year. If we ever got a blight like the potato blight in Ireland, we'd be in trouble."

Art picks up another of the farming magazines from the stack beside him. "They are storing seeds in Newfoundland if blight ever happens again. That's what would save us."

"But you couldn't put it in the ground where the blight had been," Ruth interjects, with a question in her voice.

Art says he would like to be remembered for leaving the land in better condition than it was, thus the motto of Argus Acres: "Leave it better than you found it."

Ruth would like to be remembered for the family she raised. They both hope to have passed on their good traits to their children. Art laughs.

"Don't know if they got any of the bad, too."

Any way you look at it, farming is hard work. Art's happy and engaging eyes change a little as he tells me that some people are envious of his success. Then he straightens up as he proudly states that they have always worked hard for everything they have.

Asked if they had anything to add, Art tells me one last thing:

"When I started out, all the people around me said I'd never make it. Well, they're all dead and I'm still going."

• • •

UPDATE

FEBRUARY 2015

In the years since this interview, more land has been added to the farm which now encompasses seventeen hundred acres, with five hundred more being rented.

"When we were interviewed, we said that that year had been the worst year of all, financially," Ruth tells me. "But with the higher milk prices, 2014 was our best."

A good sign for the future.

DENISE, DAVID, JASON AND GREGORY LLOYD

DAVID AND DENISE LLOYD
MAPLE DOWNS FARMS II

Middleburgh

INTERVIEWED ON APRIL 12, 2015

Along the stretch of Route 30 going from Schoharie to Middleburgh, wide, flat fields stretch out on either side of the road, while the Schoharie Creek flows gently northward on the right. This beautiful fertile valley is flanked by gradually rising hills, ending here and there in dramatic escarpments. Farms line the edge of the flat valley and rise gently up the slope, boundaries marked by lines of trees. Peaceful and pastoral, you get the feeling that nothing could ever disturb its restful ambiance. However, on August 28, 2011, the seemingly lazy Schoharie Creek became a raging tumult, and this gentle landscape was ravaged by one of the worst floods in memory.

Maple Downs Farms II, owned by David and Denise Lloyd, sits on the eastern boundary of the valley, the long driveway bordered by broad fields, the house and barns spread out just where the slope begins to rise behind them. David Lloyd was born high above this valley in the hills eight miles south of Middleburgh, on a farm called Maple Downs. He graduated from high school in 1970 and four years later was ready to go out on his own. Starting with forty-six head of cattle, he rented a farm of two hundred and thirty eight acres, one hundred and twenty of which are tillable. Spring came earlier in the valley than in the higher elevations where his family's farm was, so when he first started out his father and brother would come and help him till the fields and plant the crops. He continued to rent that farm for the next twenty-seven years and is still there, now as owner. Including the additional land rented to grow more crops, Dave now farms a total of nine hundred acres.

Denise Risse and her family moved frequently – they lived up and down the east coast from Florida to Maryland, and for a time, even overseas in Holland. Denise's grandfather, Joe Risse, ran a dairy farm in Middleburg and her family would come every summer from where ever they were to visit for a few weeks. During these yearly visits she realized she loved to farm and to milk cows.

51

Dave and Denise eventually met and married. They have three children, Christie and Jason from Dave's previous marriage, and Gregory who was born in 1992. Christie lives nearby in the village of Middleburg and has a seven-year-old son, Jakob. Jason has his own place down Route 30 near Barber's Farm. Greg lives in the upstairs apartment at the farmhouse.

In 2001, David, Denise, and Jason became partners. The following year they bought the farm that Dave had worked for so long. In a few years Greg, presently a full time employee, will also become a partner. After they purchased the farm they built twenty more stalls, and they continued to increase the herd's numbers to the present one hundred and fifty milkers. In 2003, in order to facilitate milking this many cows, they put up a free stall barn, which holds half of the cows while the other half are being milked in the dairy. When the first half has been milked, they turn them out into the pasture and bring in the others in from the free stall barn.

"It takes longer to milk this way," Denise says, "but we can milk more cows and not have the huge expense of building more dairy barns."

Four Lloyds – Dave, Denise, Jason, and Greg - work the farm together. Denise milks at 4:00 AM six days a week, even though she has a full time financial services business of her own. On Sundays she milks twice, so someone else can have a day off and she can have Saturday off.

Dave allows both Jason and Greg to take on responsibility, as he feels it is the best way for them to learn.

"Farm kids live it and they know what to do," he says.

Denise proudly tells about how capable the boys are.

"Jason was only nine when he was baling hay on his own. I would worry about him, thinking, 'He should be done by now, is he okay?' but then he would come back and everything would be fine. He was always very capable. When he was only twelve, Dave and I went to a farm show and left him in charge of the whole farm, including the milking. Both Jason and Greg can do it all. It's nice that Dave and I can go away to do our co-op work and shows and leave the farm in the hands of our boys."

The Lloyds also have two full time hired hands. Candace (Terrell) Wood has been with them for eleven years, since she was fourteen, and she works much more than forty hours a week. Jenna Sandy, who has been with them since last fall works a regular forty-hour week. The Lloyds employ some part time help as well - some SUNY Cobleskill students, and some older retired gentlemen, especially in the summer months. One reason that they have good, longstanding help is that they pay more than minimum wage.

The Lloyd's are very involved and active in the Agri-Mark, Inc. Cooperative, made up of 1200 members and the owner of Cabot Cheese.

"We prefer working with a co-op because they are made up of farmers who are in the business, so they make better decisions," Denise explains. "We travel, even to Washington, D.C., to lobby for farmers' rights. It helps to connect farmers with both the government and the public. We are able to do this because of the boys and Candace – as I said, they all work more than full time."

"We know many Congressmen and Assemblymen personally," Dave says. "We hosted a fund raiser here for Chris Gibson's reelection. Chris listens very well to our side of the story and the best thing about that day was how he spoke to the forty or fifty people that were here."

Dave also serves on the Town Board. "It's our way to affect our community and represent the farmers' needs and perspectives. The average consumer is at least three generations away from the farm, so they have no understanding of it."

The Lloyds show cows at various farm fairs around the country. At the time of this interview, both Jason and Greg are at the International Spring Show in Syracuse.

"We go to about five shows a year to show our cows," Denise tells me. "We have a registered herd and work hard to develop the herd by being part of a program to improve their genetics. Some of our cows are such good producers that we sell their embryos, and to do that we need to show our cows. We go to the New York State Fair, the All American Show in Harrisburg, Pennsylvania, and even to Madison, Wisconsin. We usually do well."

Because of the acclaim of winning and the overall classification of the many valuable calves produced, the breeding Jason has done on this farm has improved the qualities of their own cows, as well as those they sell. Together, he and a few partners scout out and buy cows they think will be good producers. Denise explains:

"Jason has a very good eye for cattle and is nationally known as a breeder and judge. One heifer has done especially well and has produced embryos to sell, as well as winning her class in the Madison, Wisconsin, show. Jason's work has helped put our farm on the map. Our cows and stud bulls are known internationally. All the offspring's names include the prefix Maple Downs."

Denise brings out some farm magazines such as <u>NY Holstein News</u>, <u>Country Folks</u>, and <u>Cowsmopolitan</u>, where they advertise upcoming sales with glossy photos of each cow that will be for sale.

"We had our first sale here at the farm in July of 2012. In a few weeks, we are having another and hope to do well. The boys are showing our cows in Syracuse tomorrow, and if they do well, it will hopefully raise the prices we can get for them at that sale.

"It takes a lot of work and people to prepare for a sale. And it is expensive. We have to hire a fitter who grooms and readies the cow, and a professional photographer. There has to be someone on the halter, one at the tail and one on each leg. We have to get the cow to step up on a box with her front feet and position them just so. Then we need a noisemaker or someone to wave a broom to get the cow's attention to get her to perk up her ears. Each photo takes at least twenty minutes.

"Shows are a lot of work, too. When we take the cows to shows that are far away, we have to stop to feed, water, and milk them on the way. It's not always easy. But it is fun."

Like many other farmers, the Lloyds always need to further supplement their income. Dave drives school bus mornings and afternoons, which is not only an extra source of income, but provides them with health insurance.

"Milk prices fluctuate so much that we need that other income," Dave says. "Last year was a good year with milk prices at a record high, but in 2009 the prices we were paid were so low that we had to borrow a lot of money to pay our bills. We have to pay all that back no matter what we get for our milk."

A tour of the barns at feeding and milking time reveals the great care the Lloyd family gives their herd. Six large fans create a wind tunnel to keep the cows cool when the weather is hot. The floor of each stall in the dairy barn is covered with a "mattress," a baffle made of shredded rubber that stays in place. There are fans in the free stall barn as well, and the cows there are provided with a bed of sand a few feet deep to create a comfortable place to lie down.

"The barns are cleaned twice a day," Denise explains. "The more comfortable we can make the cows the more they eat, and the more they produce. If a cow is uncomfortable or stressed, she won't produce as well. We do all we can do to prevent stress or sickness. We feed them well and in the free stall barn they can eat all they want - it is always available for them."

PHOTO PROVIDED BY THE LLOYD FAMILY

THE FLOOD

Inevitably, the conversation turns to the flood of Saturday, August 28, 2011.

"Dave is very good at estimating the time we have to move cows and equipment by knowing how many cubic feet of water is coming over the Gilboa Dam[5]," Denise states. "But that day during Hurricane Irene, we couldn't get any information. I went ahead and milked the cows earlier than the usual 4:00PM that day, expecting some flooding, and then came to the house to make a call to get some information, but it was already too late."

"We had twenty minutes," Dave says, shaking his head. "Twenty minutes to move everything. I could see the water coming over the field like the tide rolling in on a beach. But there wasn't time. In twenty minutes we went from dry ground to seven feet of water in the barn. Twenty minutes."

"We really lost everything," Denise says softly. "It wiped out our farm, our house, our trucks, my car, our business, my business on Main Street. Everything."

"I've been through five floods since I've been here, but the others were one or two feet deep. This was much worse," Dave continues, his eyes beginning to fill up. "Plus, we couldn't get any information. The stream behind the house flooded first, and I could see the cows swimming through it. Then the river, one mile away, just inundated everything. We lost forty-seven head, mostly small calves that just couldn't get above the water. Everything was flooded - the barn, the house, the equipment. The garage collapsed."

---

5 The Gilboa Dam is situated on the Schoharie Creek at the southern end of Schoharie County. The reservoir provides drinking water to New York City.

Denise points to the watermark on the front door of the house.

"There was four and a half feet of water downstairs in the house. It happened so fast I couldn't get back out to the barn and Dave couldn't get to the house. He, Greg, and our hired hand at the time, spent the night in the haymow. We had cellphones, but Dave's battery was weak because he had made so many calls to get information about what was happening at the dam, so we had trouble communicating. Earlier, a woman from the village who been told to evacuate, had spent a few hours in the woods up behind us with her teenage son. Soaking wet, they had come to the barn while we were milking and asked if they could come in to get dry. I told them to just go in the house and where to get towels to dry themselves off, and to help themselves to coffee. Funny thing: I asked them to please take their shoes off, because I had just cleaned the house!

"When I got back to the house to make that phone call after milking, the water started coming in and we all went upstairs to the apartment. Strangely, though the downstairs was flooded, we still had electricity upstairs, but no hot water. The woman and her son expected to be rescued, but my brother – from Georgia! - called and said he heard on the news that the current was too strong and the Hover Craft wasn't coming. He knew that in Georgia, but we were never told. The poor woman was so upset. When she had been told to evacuate, she had put all of her important papers in a backpack and strapped it on. Crossing the creek to get to our place, the current was so strong it took the backpack right off her back and carried it downstream. She never saw it again.

"Jason was away at a fair at the time. We had a bumper corn crop that year, and had expected to start harvesting the following day after he got back, but we lost that, too. And all the feed that we had harvested and stored got washed down river.

"By law, when you have a mortgage you need to have flood insurance. It's very expensive. I would never have written that check for flood insurance if it hadn't been a law. But I am so glad now that we had that insurance. We got no help from FEMA, some help from the federal and state governments, but without that insurance, I don't know where we'd be.

"And so many people helped us out. I don't know how we would have fed the cows and milked them without help from others. Crewells, whose farm is across the valley and on higher ground, let us put our cows in their barn and milk them there for two or three weeks while we cleaned up our barn."

"All our equipment was damaged," Dave continues. "The milking machines were destroyed. The old tractors we could repair, but the two new ones, those with computers, were a total loss and needed to be replaced. We needed to be very resourceful. As farmers, we did what we had to do. When you're a farmer, you've always been self-sufficient; you don't wait for others to do for you."

Denise goes on. "In addition to the equipment and buildings, our crops were destroyed. The cornfield was a real battle; the equipment we used to clean it up kept getting plugged up and it took so much time and energy removing it all. And, though it seems hard to believe, the weight of the water compacted the soil in the fields. The next spring it took a lot of extra work and time and fuel to till the ground because it was so hard. We had to plow deeper than usual to loosen it up enough to plant new seed.

"I never go anywhere without my makeup, but I sure went without it for the next few weeks," Denise said, smiling for the first time since talking about the flood. "For the next six weeks we had to go all the way to Jason's, five miles away, to shower. It was nice when we had hot water at home again. We finally moved back downstairs in May of 2012."

Asked if they have recovered fully, Dave's answer is straight to the point:

"There are so many ups and downs in farming that it is hard to say where we are. This year milk prices are down again, 30-35% since October, but expenses continue to go up, and we still have to pay for everything. Huge cycles are always occurring, and I guess this is one of them."

What makes the Lloyds keep going?

"We have a close knit family and are just glad everyone is alright and still here," Denise points out. "It helps that we all love milking cows. Dairy animals are fascinating. If you take care of your cows they produce so much. The miracle of birth is so wonderful. We all have a passion for the cows and for being stewards of the land. I love the land and always wanted to raise my family on a farm because of the work ethic involved and being with your family all the time, even if that is a double-edged sword sometimes. But it is great to be flexible and independent. I was always able to go to Greg's basketball games when he was in school."

Dave agrees. "I'll keep going. I am proud that both sons are working the farm with me and want to keep on farming. I like the different seasons, and doing different things. I like the fieldwork - you plant corn and see it come up - it's very satisfying. And I like harvesting. And I really enjoy being part of the Co-op. Well, I like the whole thing."

So, what makes farmers keep going?

They just do.

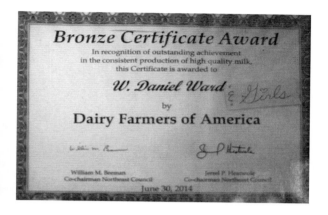

RICHARD AND JOAN PALMER

RICHARD AND JOAN PALMER

Cobleskill

INTERVIEWED ON APRIL 1, 2009

Richard Palmer can trace his Schoharie County roots back to 1880. His paternal grandfather was born in Jefferson in that year, the other three grandparents in the following decade, all within the county. Clifford Palmer, Richard's father, was born in 1912, but as a young man he left Schoharie County to take a job on a vegetable farm in Schenectady. In 1934, after only a few years away, he decided to return to the county of his birth to work on a dairy farm in Cobleskill. He settled his family, which included one-year-old Richard, in the tenant house on the Nellie Gordon Farm. He later went on to become a sharecropper on several other farms in the area - in Carlisle, Hyndsville, and Seward. He worked hard and saved his money. In 1943, he purchased the Delbert Beaver farm on Lawyersville Road.

So Richard, known to most everyone as Dick, and his younger sisters, Beverly and Regina, were introduced to farming at an early age. He says his parents were 'real workers' and taught them to be, too. Together, the family milked their eight or nine cows by hand and worked the fields with horse-drawn equipment. Slowly, they built up the herd until they had between forty and fifty milkers, and then bought a much-needed milking machine.

Eventually, they put in a bulk tank and then pipelines, but - despite the hard work of moving the heavy cans - Dick still fondly remembers the early days when they hauled the milk cans to the Schoharie County Cooperative Dairy near where the Feed Bag is now.

"The three or four hundred farmers that were part of the co-op then has dwindled to less than 50," Dick says sadly, shaking his head.

In those early years, hay was put in loose and by hand, pitched up on the wagon and then into the mow with a hayfork. Sometime in the 1950s the Palmers bought a baler, shortening the time needed for the job and making it much easier. Over the years they made many repairs to the barn, built machine sheds, heifer barns, and new silos, and slowly accumulated modern machinery.

Besides working in the fields and with the animals, Dick says you have to know how to do a lot of different things on a farm - you have to be a nutritionist, mechanic, veterinarian, and a butcher. His father was good at butchering and they would butcher 100 - 120 pigs every year for extra money.

"You'd boil water in a big iron kettle and dunk the pig in the scouring trough till all the bristles came off. When we sold the farm we left a lot of that old stuff there that I wish I still had. Some of those old implements are at the Historical Treasures building at the Fairground, now."

Dick recalls when he received his first pig.

"It was given to me by Julius Gordon, who made me promise to give him two pigs from her first litter as payment. When the pig grew up and had her litter, she had seventeen. That was too many for the mother to care for, so we brought one into the house and bottle-fed it."

After giving the promised two piglets to Mr. Gordon, Dick made enough money on the others to buy his first Holstein cow. When he took her to 4H and showed her in the ring, she won several prizes and eventually became registered, which only happens to the very best producing cows. Dick won many awards in both 4H and in Future Farmers of America, and even had a champion when he was still in High School. One year he won a Guernsey heifer in a judging contest, and another time a Holstein calf for getting the most people to sign up for the Holstein Club the following year.

"That one didn't amount to much, though," he says. "But I learned a great deal about dairy farming in FFA and 4H. I was the second recipient of the Empire Farmer Award in Cobleskill High School. And through 4H, I met Governor Dewey, and had my picture taken with him."

Dick learned other kinds of lessons about life on the farm, too. When he was twelve years old he broke his arm cranking a gas engine. Several weeks later, though his arm was still weak, he was driving the horses and pulling an old lumber wagon with a heavy load of stones they had cleared from a field. Going through a big mud puddle, the horses suddenly took off running. Dick was not strong enough to get them stopped and they kept going straight for the barn. One horse ran into the barn, the other did not, stopping their flight abruptly.

"My mom was watching and she sure was frightened, but I was okay."

Dick didn't miss the horses once they changed over to using tractors. However, he had a close call on a tractor, too. He recalls one day when he was bringing back a full wagonload of chopped corn.

"I should have come down sideways, like usual," he says, "but instead I came straight down. The load was too heavy and the John Deere tractor started sliding.

I stood up and pressed down on the tractor wheels with both feet but I couldn't stop it and it just kept sliding. The connecting pin came out of the drawbar and the tongue dug into the ground. That finally stopped it and I was okay. My father saw it all happen and he was scared, but not angry - he was just happy I wasn't hurt. You have to think fast, sometimes."

Dick's father had purchased a second 75-acre farm on Gardnersville Road three or four miles away, for the use of the land, from which they sold surplus hay for extra cash.

"There was a huge old elm tree on that property that died in the 1960s due to the Dutch Elm epidemic. It was the biggest elm in New York State, measuring twenty-four feet eight inches around. A special chain saw had to be made to cut it down. Much of it was sold to individuals, but one big piece is at the Stone Fort in Schoharie. We made the TV news!" he says.

Dick's wife, Joan, although not a country girl herself, was familiar with farming from visiting her grandfather's farm in Clinton when she was a child.

"But I was really a city girl."

When they met, Joan was a young widow with two children, Lori and Ken, ages 11 and 6. When she and Dick married in 1974, they combined their families – he had two children, Brenda and Dan, ages 14 and 12. It was that same year that they bought the farm from Dick's father.

Joan took right to farm life. "I had to learn there were some clothes I shouldn't wear in the barn, but otherwise I loved it. I was amazed to see that each of the cows knew just which stanchion was theirs. They would never go into a different one. Our children were involved in 4H and showed cattle, and Dick and I were both leaders for while. I also did the bookwork for the farm."

Just as Dick had done on his father's farm, at first the family did most of the farm work by themselves. During the summers they would hire high school kids to help with haying and as the farm continued to grow they did eventually employ a hired hand.

Asked about what was the worst thing about farming, the couple answered in unison…

"The hours and the winters!"

"It was terrible when it was 25 degrees below zero and all the buckets froze," Dick explains. "Those were tough days. But summer days could be tough, too. One hot day I was leaning against a steel pole and milking a cow. There was a thunderstorm going on and lightning struck the barn and came down the pole. I swear I saw sparks come out the ends of my fingers. Most people don't believe me, but I swear it's true."

Joan tells of another summer day when the cows got out and were running down Loonenburg Turnpike.

"They got into the corn field and we couldn't see them. It was like a maze, and it took five of us an hour to round them all up and get them to the barn."

When they sold the farm to Ron and Becky Gage in 1989 they were milking fifty to fifty-five cows and had thirty-five young cattle. They are very happy that their farm continues to be a dairy.

"Today's farmers have it rough," Dick says. "It's hard to make it today. The price they get for milk goes down and the price of feed and everything else they need goes up and up. It's hard to afford to stay in farming today. I feel real bad about how many farms are being sold and subdivided."

After he retired, Dick sold minerals, a necessary additive to cows' feed, to other farmers. He gave it up after just a short time because he felt that most of the farmers really couldn't afford them.

Dick would like to be remembered for being the recipient of several production awards, including the top Dairy Herd Improvement Award, and for being the Director of the Eastern Artificial Inseminators for twenty-five years. He sat on the town board and for eight years served as Town Justice.

"I feel I was a good and fair justice, and I performed a few weddings, too. Sometimes, if it happened to be milking time, I would see those being arrested right in the barn. There were lots of different cases, but I especially remember when five young men were brought in who had been driving their car on the railroad tracks and jumped out when a train came, leaving the car on the tracks."

Other times the justice work was done in the house. Joan remembers an incident that involved two young girls who had been caught smoking pot.

"I could hear them giggling all the way upstairs," she laughed.

Dick feels his biggest accomplishment was that he was a success. "I always kept my machinery in good condition, and never bought any piece of machinery I couldn't afford. If I couldn't pay for it, I would wait to buy it until I could."

But their life has not been all work. They reminisce fondly about the Pumpkin Hollow Boys (and Chicks!), a fun club that was started fifty years ago by Loren Guernsey, another diary farmer.

"It was just a group of people who would get together and build parade floats of all sorts, on old lawnmower parts. We were in parades all over Schoharie County. We had lots of fun," explains Joan. "We are also active members of an Antique Auto Club and we own a 1953 Chevy."

When they sold the farm, Dick and Joan built a new house on part of the land that they kept. They also kept a piece of the farm that has a pond, where they have a little family campsite and have picnics.

Dick and Joan regret that none of their children or grandchildren live nearby. They say that their grandkids are the "greatest grandkids ever," a sentiment that is apparently reciprocal, as they both wear a lapel button that says, "The Greatest Grandparents Ever."

Brenda and Dan both live in Rochester. Brenda and her husband, Greg, have two children, Ryan and Jordan. Dan and his wife, Patty, have Stephanie and Jonathan. Lori, and her husband Dave, live in Greenwood, Illinois and have three children, Luke, Zachary and Angela. Kenneth lives in Clearwater, Florida, where he enjoys the single life.

The Palmers continue to stay busy. Joan is on a bowling team and bowls every week. Dick still mows the spacious lawns around their house, and until very recently, also bowled every Tuesday night. They are both active members of the Lawyersville Reformed Church. The final thing that he mentioned when listing his accomplishments is that several years ago he celebrated being a member of that church for fifty years.

"It doesn't have anything to do with farming, but I joined that church as a teenager and I'm proud of still being an active member."

FOUR GENERATIONS - *L TO R:* DAN, JONATHAN, DICK AND CLIFFORD PALMER
PHOTO PROVIDED BY PALMER FAMILY

ABOVE: ELMER AND SHELMA ROCKWELL
BELOW: ALICIA, HER DAD TIM, AND BROTHER MIKE

ELMER AND SHELMA ROCKWELL

Cobleskill

INTERVIEWED ON NOVEMBER 7, 2014

A collection of unusual rocks sits in one corner of the small flower garden that borders the deck leading into the Rockwell house. When Elmer works in the fields he likes to collect rocks that remind him of other things. He points out a large horse's head, and a "stone-age shoe." Farther out in the garden is one he calls The Flintstone Car. Elmer Rockwell is a man of great imagination.

Elmer was born in Howe's Cave on December 29, 1935, the fifth of ten children and the first boy. Soon after his birth, his parents moved to a farm on Sagendorf Corners, and when he was seven they moved to what is now the Steidl farm on Hill Road in Cobleskill, then owned by his maternal grandfather, Lambert Nethaway. He remembers his parents farming with teams of horses, and cutting ice in the winter to cool the milk in the summer. Elmer loves to talk about his family, his memories, farming, and work ethics.

Shelma Borst, born March 25, 1938, grew up in the village of Seward, attended the one-room schoolhouse there, and graduated from Cobleskill High School. Her dad was a mechanic and carpenter, and excelled at fixing John Deere tractors. During high school Shelma worked as a cashier for the A&P, a job she loved. She later worked as bookkeeper for Selkirk's Hardware, which was located where A Cut Above is now.

Elmer and Shelma started dating when they were in high school and at one time she started a rumor that they were going steady. The night that Elmer heard the rumor, he formally asked her and she readily agreed.

"All of our friends had already been chanting a poem about us: 'Shelma Borst loves the cows. She's helping Elmer in the mow,' " Shelma says, smiling. "We got married on May 12, 1957, and after that I finally 'fessed up that I was the one who had started the rumor."

Elmer and Shelma have four children - Sherrie Bell lives in the apartment in the house where Elmer and Shelma lived when they first moved here. Lisa Bly lives in Berlin, NY. Tim, a mechanic and welder, lives on the adjoining farm where his daughter now milks cows in the same barn her granddad did. Shawn is a Commander in the US Navy. They also have eleven grandchildren and five great-grandchildren.

Elmer worked on several farms before buying his own, including the Nellie Gordon farm on Route 145 and Corbin Holmes's farm on Mineral Springs Road.

"I was very good friends with Corbin Holmes, who never had any children. He gave me a lot of advice and taught me a lot of things. He offered to leave me his farm in his will, but I turned that offer down. I knew kids who had been given things for nothing and they never turned out well. I believe you have to work for what you get in order for it to be meaningful.

"Corb helped me a lot. He taught me to look carefully at the soil on each farm. Some farms have good soil but haven't been managed well. Others might look good, but don't really have good soil. You need to know what to look for."

In the end, it was Shelma that chose the farm. "When we were going steady, we drove down this road toward his family's place and I liked the looks of this house. I said, 'Elmer, buy me this farm.' And about ten years after we got married, he did."

Elmer smiles at the memory. "While I was working at the farm on Mineral Springs Road I always dreamed about buying my own. But I didn't think I'd ever be able to."

In 1966, however, their dreams came true, and they bought the farm on Hill Road that Shelma had long wished for.

"We had nothing but a dream, so we had nothing to lose," Elmer says. "We moved here in February of 1967. We both worked really hard. I loved every minute of it. Eventually, we also bought the one hundred and thirty acre farm that adjoins our place and moved the cows over there. It's in that barn that our granddaughter, Alicia, milks now. She milks about forty-five cows and has about thirty-five heifers. Her dad and her brothers, Kenny and Mike, help her out."

Elmer talks about how farming has changed over the years and how hard it is to hire good help these days.

"I used to be able to hire boys to help with the work. I could fill the whole barn with hay with three hired boys. I would rib them a little and make a bet with who could move the most hay. I would bet them an ice-cream cone and then make sure I'd lose. So I would treat them all to ice cream when we were done.

"We used to bale square bales, which you could handle by hand. You see the white wrap there? And there?" He points out one window, then another. "You can't get the labor anymore; they won't work. They won't. So you need to put green hay in plastic and handle it with tractors.

"Boys used to love to carry the calves and clean out the mangers. I'd have to supervise them, but they loved doing it. You used to be able to get boys to work their hearts out." Elmer shakes his head. "You can't get boys to work like that anymore."

"Nope. He's right," Shelma adds. "They have no sense of urgency or enthusiasm for it."

"We've worked hard but loved it," Elmer says again. "And I know a good idea when I hear one and I can usually figure things out. When we lived on the Mineral Springs farm I wanted a snowplow. I kept thinking about it and trying to build it, but had no money. One morning I woke up and told Shelma that I knew just how to do it. I just dreamed it. I built that plow for $10 and plowed driveways up and down the road, made $10 on every driveway. The Town Barn let me have the chains they wore out on the highway. They were good enough for the tractor to do driveways."

Like many other farmers, Elmer and Shelma think that the farm is the best place to raise a family.

"I don't think I ever spanked my kids," Elmer said. "But discipline is necessary. We tried to show them the right way by example. They liked to camp in the woods and I showed them how to safely start a campfire. I'd tell them, 'Use your head.' We never had a problem."

"Well, except they took all my Tupperware," Shelma interjects.

"Once the neighbor's kids built a fire under the hemlock trees and started four fires. We ran over and put the fires out and I overheard one of my kids say to his friend, 'I told you my Dad would be mad.' And I was."

"Our kids learned the hard way sometimes, but they learned," Shelma says. "One time Sherrie didn't want to baby sit for Shawn so we hired a baby sitter. Boy, she was mad. When we got home she said, 'Don't you ever do that again! I can watch my own baby brother!' " Shelma and Elmer both laugh at the memory and it triggers another for Elmer.

"Another time Shelma and I were going to go shopping and the girls were to watch Shawn. He cried to go along, so we took him, but we forgot to tell the girls. They were scared to death 'cause they couldn't find him. When we got home we realized what we had done. We knew we were wrong and we admitted it. If parents can admit they're wrong, then the kids will learn to do it, too."

"Some parents today are afraid to discipline their children," Shelma mused. "But discipline doesn't mean you have to beat them up. You just correct them and set a good example. The kids loved to go to Brooks' BBQ, but they knew they had to behave, or we wouldn't take them again. It was simple."

"Sometimes they would get mad and have a fight, but we mostly let them work it out themselves," Elmer says. "I remember once having a fight with Leonard Holmes…he was a few years older than me and while we were playing he grabbed some pigeon 'droppings' (not what Elmer said) and stuffed them down my neck. That made me mad and we got in a fight and I went home. When I got there, my father was on the phone with Leonard's dad. I expected to get in trouble, but then I heard my father tell Mr. Holmes, 'If we interfere they won't be friends any more, so we should let them work it out themselves.' And we did, and we remained friends. My father was a smart man that way."

"Our kids had a good time on the farm," Shelma says. "And they appreciated it, even the discipline. Once they wanted to go snowmobiling and I told them to be home early. When they came home at 4:00 am, I told them I had meant for them to come early in the evening, not the morning; we were worried sick. One of their friends said 'Thanks for caring about us,' but we didn't let it go. Elmer made them milk the cows and he went back to bed."

When asked what they liked best about farm life, Elmer had no hesitation: "I can be my own boss, use my brain, make my own decisions. We're close to nature and can look around us and enjoy our surroundings."

Shelma feels that the hardest part of farming was having to be there every day to milk the cows and do the chores, never having a day off.

"But when the kids got older and could do the milking, then we could go away a little."

Elmer agrees. "There were hard times, lots of them. One of the worst was when Lisa was about five years old, she came running to me yelling, 'Daddy! There's milk running out in a stream.' I went running over to the barn in my bare feet. It turned out that a neighbor kid had opened the spigot to the milk tank and couldn't shut it. Before I could get it closed, we lost a few hundred pounds of milk. That was a lot of money."

Elmer feels that their greatest accomplishment was raising four kids. "We had a close relationship with our kids. We'd tuck them in at night when they were little and always spent time with them. We want to be remembered for being good parents."

"There's nothing more important than that," Shelma adds. "And to set a good example and try to live a good life. But it's also important to learn not to let people take advantage of you. Once a man who helped Elmer liked me and he would chase me around. I figured I was younger and thinner than him and I could get away from him, but one day while I was sweeping, he cornered me and said, 'Ahah! I gotcha now!' 'No you don't,' I said, and hit him with the broom. He never came back." They both laugh. "Life gets hard as you get older, though. We're both seriously diabetic now, and Elmer has fallen a few times."

What would the Rockwells like to be remembered for? Elmer is quick to answer: "Being a hard worker and knowing how to use my head. Like I said, when we bought this place we had nothing but a dream, so we had nothing to lose. But we worked hard and learned to make the best of what we have. That's one thing I don't think the second generation does so well. I think they are too hovered over."

Elmer also offers some advice: "Build a good fence, so your cows don't get out."

Shelma agrees. "It's a big nuisance when they do."

"Fences are important," Elmer says. "A good line fence makes very good neighbor."

It appears that Elmer and Shelma Rockwell were very good neighbors.

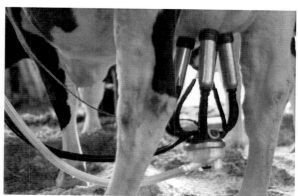

73

LAKEN, DEBBIE, JOHN, ERIC, CADEM STANTON

JOHN STANTON
STAN-AYR AND STAN-HOL FARM

Cobleskill

INTERVIEWED ON FEBRUARY 20, 2013

Farming is hard on the body. Take it from John Stanton, who is recovering from a hip replacement, his third surgery over the past few years.

John was born and raised on his dad's dairy farm in East Cobleskill. He says he has farming in his blood and a passion for the cows. After graduating from Cobleskill High School, he attended SUNY Cobleskill for two years while also working on the family farm. He then went to work as manager and herdsman for Philip Schuyler, a well-known and respected area dairy farmer.

"I learned from a good one," he says of Philip Schuyler, "and always appreciated it."

He stayed in this position for fifteen years, the last two of those years managing the farm of four hundred cows alone, as Mr. Schuyler's health was deteriorating.

"He had confidence in me," John says.

In 1991 John took over the farm, leasing it through a private deal but buying his own equipment. Ten years later came the real estate deal, and the farm was his.

"This is the only way young people can get into farming today - that or inheriting it from their father. It's just way too expensive for them to start all by themselves."

Debbie Felter also grew up in Cobleskill, graduating two years behind John from both high school and SUNY Cobleskill. They married in June of 1976.

"She was my high school sweetheart, I guess you would say."

Debbie does the bookkeeping for their farm and has always worked in banking and in real estate, as well.

John and Debbie have one son, Eric, 29, who also graduated from SUNY Cobleskill. He is now a partner in the farm and is working toward eventual ownership. While John continues to recover from the surgery he had in early January, Eric is running the farm alone. Eric and his wife, Laken (Baker), and Eric's three-year old son, Cadem, live next door in the big farmhouse. They are teaching Cadem to feed the calves and he loves doing it.

"He's amazing," John says, smiling. "Already, he can look around and tell what's different." Through Cadem, John sees a future for his farm and hopes it will continue to thrive.

"You are always hoping for the best. Growing up on a farm gives kids a good start for their future. A farm will keep a kid out of trouble - I don't know of too many farm kids that get in trouble. The hours are long, especially in the summer - it teaches them strength and perseverance. Farming gives you a great sense of accomplishment. It's a hard life. It's a good life."

John is also a dairy coach, knowledge he learned from an Ag teacher many years ago. He once made a gift of a calf to a local girl who learned how to care for it, and she grew up responsible enough to do milking for him when he needed someone to fill in.

"She'll be something someday. She learned a lot from caring for that calf, including perseverance. I've seen a lot of farm kids grow up and do great things. The farm, for me, is the place to bring up kids. Keeping the family intact is very important. And so is putting your best foot forward and being grateful that you've gotten this far."

As a 4-H leader, John has watched many young people grow up and become responsible adults, which has given him a great sense of satisfaction. In over thirty years in 4H he has touched a lot of people, and one of his frequent pieces of advice is to join farm organizations in order to meet a lot of great people, open doors to other avenues, and learn from others. He feels that being in 4-H was the fun part of being a farmer. On trips to visit different farms with his 4-H members he found there was something to learn from every farmer.

John sees a tough future for farmers. He says that over the last six or seven years it has been especially difficult.

"There have been lower lows and the highs have not been high enough. But I am eternally optimistic."

John and Eric have a total herd of four hundred and twenty cattle, two hundred of which are milkers. They are always working to get larger, in order to make more money to get by.

"It's a numbers game," he explains, "and farmers are at the bottom of the totem pole. Taxes are always getting higher; things always cost us more. There are too many restrictions and regulations, and we are the most stringently inspected industry. There are too many middlemen, and the price we are paid is determined not by us, but by the government.

"Dairy farming is the #1 industry, bringing in the most revenue. When we pay our bills the money goes out to so many businesses. The milk cooperative I belong to employs 800-1000 people. There are many, many farm-related jobs. Think of it as an umbrella. We are few in people numbers, but big in monetary numbers. We get frustrated when politicians sit on their duffs and don't get decisions made."

John is concerned about that statement being published, but it is a recurring theme among all the farmers; he is not the first farmer to say it.

And yet…. John loves the daily challenges and the variety of things that must be done on a farm. He loves the change of seasons and what comes with each season, putting seed in the ground in the spring and seeing your crops grow, hoping everything will go as planned and hoping you'll make it through.

"You live on anticipation. There are triumphs and there are troubles - bad decisions, tractors blowing up. Then when you get your crops all in on time and see what you've accomplished, you say, 'Wow. How did we do that?' No matter what kind of farming you are talking about, putting seed in the ground and watching it grow is the best."

That is another common theme among farmers – the hardships *are* hard but worth it. The financial stress is also emphasized over and over again. Farmers continually ask themselves - why can't we do better? John points out that farmers have a lot of money lying out there.

"This past year was better, but still not enough to give us any financial security. And now they want to increase the minimum wage. What they don't understand is that we just can't afford to pay more. I'm not convinced that it will really help the general economy any, and it *will* hurt us.

"Years ago, we could hire high school and college kids, but today they just don't want to work. So we have recently hired two men from Guatemala through the Western New York Agency. Guatemalans do a lot of their own networking to find employment. They know how to work and they work hard. They want more and more hours and aren't happy unless they are working 70-80 hours a week. With this extra help we can milk the cows three times a day, boosting milk production to bring in more income. If you feed the cows good enough they'll produce."

One of the things necessary to learn on a farm is caution. John talks about many close calls over the years, including two recent episodes. Not long ago, while spreading manure on a steep hill, the tractor ran away with Eric, careening down out of control. Fortunately, Eric was not hurt and there was only minor damage done to the equipment. In another near catastrophe, the tractor John was driving near one of the storage bunkers must have gotten too close to the edge; it suddenly rolled sideways, tipping over with John landing underneath. Eric and others who witnessed it called 911.

"Everything stopped. A man who happened to be driving by left his truck running right in the middle of the road to dash to my aid. I crawled out from underneath the tractor, unhurt except for a gash on my head. I sat on the ground feeling a little delirious for a bit, and then yelled, 'Would someone please turn my tractor off?' God must be with us. There's a scary dark side to farming. There are too many close calls. You need four to six eyes in your head to watch everything."

John turns the conversation back to more positive things. There are many good memories of showing cows at various fairs when John was younger, though now he only shows locally.

"In our heyday, when Eric was only fifteen, he had an Eastern National Grand Champion Ayrshire at the Eastern States Expo in Springfield, Mass. That cow went on to be nominated for All American, one of the top 5 or 6 Ayrshires in the country.

"Both Eric and I can distinguish good cattle. The cows come first. The equipment comes as a need to care of the cattle. We practice all the time in order to get better." John smiles. "You get out of it what you put into it.

"In the 'good old days' you had no barn cleaner, you pitched hay by hand, carried the milk from the cows to the milk house. But today, in order to make a living, you need to get bigger. Getting bigger means more equipment, which costs a lot more money. A used baler recently cost us $75,000. And that's just one piece of equipment."

John doesn't like round bales because of the waste when they sit outside.

"But if you prefer to put in square bales, you need more space. We don't have enough storage space for our crops. We're always upgrading. My wife complains we always want more. But we're always building for another year."

John picks up a farming magazine and taps his finger on the cover picture - a field of tall green grass.

"Green grass is what we want to see. In years of drought, such as 2012, I worry a lot, but we run enough land that we get by. We had enough to feed the cows."

Planting the crops and seeing them grow. Bringing in the harvest. These are essential, and the farmers have an emotional tie to the land and to the crops that grow on that land. John mentioned the 2013 Super Bowl advertisement by Dodge that featured farmers, and how much it meant to him. His eyes fill up as he talks about it.

"I'm choking up just thinking about it. It was so well put together, and Paul Harvey's voice and words were just perfect." He stops for a moment to wipe his eyes. "Farmers like to be appreciated, and I sure appreciated seeing that ad."

As he gets older John sometimes thinks about doing something easier, like selling or delivering parts. But he's also thinking about how to keep encouraging Eric when *he* gets discouraged. He wants to keep Eric involved until the time comes to turn things over to him.

John's cell phone rings. It's Eric, needing John to track down a part for a vital piece of equipment. Even when recuperating from surgery, a farmer has work to be done.

• • •

UPDATE

FEBRUARY 2015

A granddaughter, Aubree Stanton, was born on her parents' wedding anniversary, August 11, 2014. Though she had a difficult birth and was hospitalized for the first few weeks of her life, she is now thriving. John's hip has healed well.

AT LEFT: **DAN WARD**

BELOW, L TO R: **KAYLA, DAN, SASHA, DANIELLE,**
FRONT ROW: **KATRINA**

DAN AND GINGER WARD (AND GIRLS!)

Sharon Springs

INTERVIEWED MARCH 5, 2013

Dan Ward had a dream. As a child growing up on his family's dairy farm in Carlisle he dreamed of someday running that farm, which had been in the family for three generations. In his dream he had a wife and a close-knit family, all of them working together, bringing the farm into the future still in the Ward family.

"I always wanted to be a farmer; I like the cows and it's all I've ever known."

However, Dan's dream was not to be. His oldest brother inherited his parents' farm, and unlike Dan, he was not interested in continuing it. And so, the family farm fell fallow.

But through determination and hard work, Dan did make the rest of his dream come true. In August of 1992, he married Ginger Madl from Howe´s Cave, whose family had a hobby farm with goats and pigs and cows and lots of other animals. Ginger also had a dream:

"I wanted to be a veterinarian, but instead, I fell in love with a farmer." The Wards have rented the Donnie Lambert farm on Argusville Road for eighteen years.

Dan and Ginger have five daughters: Tanya, 19, Kayla, 17, Sasha, 13, Danielle, 10, and Katrina, 7. Katrina was born during the devastating hurricane of 2006, thus her nickname, "Hurricane," which her family says suits her well.

"People sometimes say 'too bad you don't have any boys,' " says Ginger. "Well, I don't need boys. These girls can do anything."

The girls work on the farm right along with their parents. Along with Dan, Ginger and Danielle milk at 5:00 AM and Sasha and Kayla do the evening milking. Dan says that if he had to hire anyone he wouldn't be in business at all, because of the cost. But he is quick to go on.

"A farm is a great place for kids. And farming teaches the kids that they have to work for whatever they have. They know they've earned it, and can be proud of that. And they learn to fend for themselves. I'm proud of what I have accomplished, and that I built up everything on my own. I hope to be remembered for always working hard. But there's a down side, too - we never get to take a break. Even our honeymoon was interrupted because the bulk tank broke down while someone else was milking for us, so we had to come home to fix it. The only thing that could make me quit farming would be if I ever had health issues. Then maybe I'd think about doing something else. But I like the animals and I have a lot of great memories. One hard memory is of losing my family's farm because of the belief that the oldest boy should inherit the farm. And I hate the paperwork, and I don't like the finances. I don't like that people make us do so much paperwork. That's why I won't borrow money, because they want too much paperwork done."

Like many farmers, Dan has to find ways to supplement their income. He is a good mechanic and does heavy equipment and tractor repair.

"We also sell hay," Ginger says. "We can't just sell milk because there's too much outgo and not enough income."

All the girls say that they love the life that farming brings, and they are eager to share their thoughts. Big sister Kayla does have concerns about her little sister's safety in the barn, as they all have had their share of mishaps.

"I don't like Katrina to be in the barn and around the cows," Kayla says. "I'm so afraid she'll get hurt. There have been so many close calls. But my worst fright was when a bushhog Dad was repairing fell on him, crushing his chest. He was screaming because it was crushing his ribs. I lifted it off of him. I don't know how I did it because it was so heavy, but I did. My knees were actually knocking against each other I was so scared."

The girls all do have scars, mostly from close encounters with barbed wire. But one of their biggest scares was when Katrina ran out onto the manure pile to gather one of her chicken's eggs.

"She started to sink, and I ran out to get her and we both were sloshing around and we sure did stink," says Ginger. "The smell was awful, and I had to hose her and myself down with cold water before we could come in and get cleaned up. But the worst of it is, that egg had been thrown out there because it was rotten. Katrina didn't know that."

Tanya graduated from Sharon Springs High School in 2010 and now works in construction as a heavy equipment and excavator operator. Kayla, who will graduate this coming June, is thinking of following her sister into the construction business. Of all the girls, Danielle is the only one who is interested in farming as a life's work.

"It's too hard, too much work, and you always get less than it costs," the older girls all say.

But Danielle is quick to disagree. "I don't think the work's so hard. I still want to farm when I grow up. I like it. Maybe that's why they call me the 'boy' of the family."

Kayla jumps in with enthusiasm. "Farming is a whole different experience. It's interesting and fun. You get to ride snowmobiles and there's always something to do, so you're not just sitting around all the time, like some of my friends in school."

The older girls enjoy knowing many things that their non-farming schoolmates do not get to learn.

"I have a friend who is from the city who moved here a few years ago to a farm that raises beef cattle," says Tanya. "He didn't know anything. He thought he'd have to milk the cows. I had to explain that you don't milk beef cattle. I like that we know all these things. Kids that live in cities should visit farms, so they can see how we work for what we have. They need to know that money isn't everything - that you can really live on very little money. Best of all, you have each other. When you get stressed, you yell at each other, but then it goes on and everything is okay."

Katrina loves her animals. "I have goats and baby pigs and calves. The goat likes to play with me and I get to pick him up and he gives me kisses. My favorites are the calves. They chase me around and I get to play with them and feed them."

Dan and Ginger agree with the rest of the dairy farming community that there is a big disconnect between farmers and the public, including government officials.

Ginger explains. "People need to know where their food comes from. Milk, eggs, cheese, yogurt – it all comes from our farmers. When you are a farmer, you work on something that's yours. Everything you have you've worked for - that's a good feeling. And I never needed to put the girls in day care. The things I don't like are the bills, finances. There's not enough income for the outgo, which is why Dan is trying to expand. I'm glad for Dan that he's got his kids with him on a farm. That's what he's always wanted, a close family, and that's what we have."

In order to increase milk production and raise their income, Dan and Ginger have recently rented Ray Briggs's barn on Rock District Road, a five or six mile drive from their home farm. This barn is bigger and can hold more cows than the present forty they are milking, and has room to expand the herd. They are using their original barn for the calves and heifers.

It is evident that this is a close family. The girls are gentle and caring with each other, even though there was a fair amount of good-natured teasing as they shared their after-school snack of yogurt and cold pizza. There is a comfortable camaraderie between them and their parents that is unmistakable. There is a good feeling there. And love. And pride.

And hope for the future.

• • •

UPDATE

FEBRUARY 2015

Dan has followed his plan of increasing his herd and is now milking sixty cows. In order to increase milk production he has also started to milk three times a day. Since Kayla graduated from high school she has been working for a nearby ARC home and helps her Dad milk at noon, before her afternoon shift there begins.

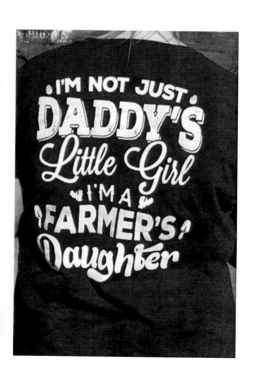

META WATTS
Author

Meta Watts is a musician, Registered Nurse, and Certified Healing Touch Practitioner, who lives in the Town of Seward, just north of Cobleskill. Born in Gardiner, New York, Meta learned how to milk a cow and pluck a chicken nearly before she learned to read. Her drivers' education was done on the tractor while working in the fields.

Meta and her husband, George, who grew up on a dairy farm in Gardiner, moved to Cobleskill in 2002, happy to be back in dairy country, since there are no farms left in their hometown. Saddened by the fact that so many of the Schoharie County's farmers were going out of business, she decided to write this book in their honor.

When not engaged in their many other activities, Meta and George enjoy spending time with their three children and seven grandchildren.

METAWATTS@GMAIL.COM

BARBARA NARK
Photographer

Barbara Nark grew up on a two-family farm in Schoharie County, New York. Her childhood was filled with family and friends, hard work and fun. Life on the farm shaped who she is as an artist today.

The range of tasks necessary to run the farm taught Barbara that creativity is important to almost any medium of work. As an adult, she spent many years working in wood and stone building construction. She continued to refine her artistic instinct and creativity with this work.

On her 5th birthday, a relative gave Barbara a Kodak Brownie Instamatic camera. This was the start of her love for photography. She started shooting animals and landscapes. Later, she photographed weddings and portraits. More recently she has been photographing local artists' work for advertising and reproduction. She has also shot instruments and musicians for album covers and promotional materials. Her work has won several awards.

Being creative has always been an important part of Barbara's life. Growing up on the farm and working in many creative mediums has refined her artistic capabilities. It has instilled in her a deep artistic sensitivity. Photography and music are the dominant means she uses to explore and express her artistic sense.

Barbara looks forward to spending the next chapter of her life creating even more art and music. She sees endless possibilities to explore.

BARBARANARK@GMAIL.COM

So Long!

BARBARA NARK'S GRANDSON, FREDDRICK SPERBECK